Asia Congress on Energy, Environments and Ecology Advances and Developments 2016

Dubai, United Arab Emirates
15-16 January 2016

Editor:

Fattaneh Daneshmand Malayeri

ISBN: 978-1-5108-3259-6

Printed from e-media with permission by:

Curran Associates, Inc.
57 Morehouse Lane
Red Hook, NY 12571

Some format issues inherent in the e-media version may also appear in this print version.

Printed by Curran Associates, Inc. (2016)

For permission requests, please contact Imperial Consortium of Research and Development (ICRD) at the address below.

Imperial Consortium of Research and Development (ICRD)
P.O. Box: 116974
Dubai, United Arab Emirates

Phone: +971 45574074

info@imperialconference.com

Additional copies of this publication are available from:

Curran Associates, Inc.
57 Morehouse Lane
Red Hook, NY 12571 USA
Phone: 845-758-0400
Fax: 845-758-2633
Email: curran@proceedings.com
Web: www.proceedings.com

TABLE OF CONTENTS

Isolation and Identification of Procambarus Clarkii Bacterial Pathogens

Dongren Zhou, Xueping Ye, Yizhi Luo, Xiaoyin Hang
Zhejiang Institute of Freshwater Fisheries, Huzhou, Zhejiang, 313001

Abstract: A kind of serious disease occurred in the cultivation of procambarus clarkii in Anji of Zhejiang in 2013. From the isolation of bacterial strain N201307 in the intestinal tract of diseased procambarus clarkia to the artificial infection of healthy procambarus clarkia, nearly half of the amount was dead within 8 days, and the median lethal doseL D50 was 5.3×105. The infected procambarus clarkia showed up similar symptoms with the naturally diseased ones. Through Gram staining, the Meriel bacterial determination reagent strips ID32 E were identified as aeromonas hydrophila. In order to further determine the taxonomic status of pathogenic bacteria, the 16S rRNA sequence was amplified and tested, and it was fond that the homology with bacterial strain ATCC7699 reached 99% after the sequence alignment carried out in NCBI gene pool. It has been confirmed through the above mentioned experiments that the pathogeny of large-scale death of procambarus clarkii was aeromonas hydrophila. The drug sensitive test result reflected that this bacterium was sensitive to the doxycycline, chloramphenicol, enrofloxacin, Ciprofloxacin, and Rocephin.

Key words: Procambarus clarkia, aeromonas hydrophila, biochemical property, drug sensitive test

1. Introduction

Procambarus clarkii is also called red swamp crawfish, known as crawfish and attached to the Arthropoda, Crustacea, Malacost raca, Decapoda, Pleocyemata, Astacoidea, Cambaridae, Cambarinae, and Procambarus. It is the species with great economic value. In recent years, the artificial breading area is gradually enlarged due to the relatively high return on equity, but the occurrence of disease becomes the bottleneck problem of the sustainable development of the current Procambarus clarkii breeding industry[1]. In June 2013, the Procambarus clarkii aquiculture area in Anji suffered from a great disease, with the following symptoms: at the first beginning, the diseased in NCBI gene pool. It has been confirmed through the above mentioned experiments that the pathogeny of large-scale death of procambarus clarkia was faint and slow in action, the inner wall of test had no vitiligo or other symptoms [2], it could be seen from the dissection that the hepatopancreas was swelling, the intestinal tract was filled with blue blood and without food. The death rate at the initial stage was low, and the death quantity increased gradually, which was as high as 40% in the fastigium. Consequently, the local farmer suffered from tremendous economic loss. A strain of fetal aeromonas hydrophila was isolated from the diseased Procambarus clarkia, and the main biological characters, 16S rRNA gene series were studied. Moreover, physiological-biochemical characteristics analysis and drug susceptibility tests were carried out for the isolated strains, which provided bass for the effective prevention and treatment.

2. Material and method

2.1 Strain isolation

Diseased Procambarus clarkia was from Zhejiang Zhongzheng Agricultural Development Co., Ltd. The typical symptoms were that there was no food in the intestinal tract, and it was filled with blue blood, the gill turned yellow(as Fig1). In aseptic condition, the intestinal tract and hepatopancreas tissue was taken to draw lines on the nutrient agar plate and horse blood plate, and it shall be cultivated at 30℃. After 48 hours, single colony shall be taken for pure culture [3].

Figure1 The gill turned yellow

2.2 Pathogeny morphologic observation

Purified single colony was selected and cultivated for 48 hours at the temperature of 30℃, and it shall be observed under microscope after smearing, fastening and dyeing.

2.3 Challenge test of isolate for Procambarus clarkii

Healthy Procambarus clarkii was taken from a certain farmer in Anji, Huzhou for challenge test, and seven healthy Procambarus clarkii, about 8 to 10 cm were selected from each experimental group. The bacterial colony was selected and placed in Brain Heart Infusion in shake culture for 48 hours. The bacterial suspension was prepared from the stroke-physiological saline solution, and the concentration of adjusted bacterium was 1×108, 1×107, 1×106, 1×105 and 1×104. Challenge test was conducted for the healthy Procambarus clarkii in intramuscular injection in the tail with the bacterium of different concentrations, and the injected dose was 0.1 mL. After Injection, it shall be fed in the water cluster about 100cm in diameter and 10cm in depth, with a brick broaching in the middle, as well as some water hyacinth. The temperature shall be controlled at about 25℃. The morbidity shall be observed and blank control group shall be set. The morbidity symptom and death condition of the procambarus clarkii was observed and recorded, LD50 was calculated with Karber method, and bacteria re-isolation was conducted from the focus of the dying procambarus clarkii [4].

2.4 Identification of physiological-biochemical characteristic

After Gram staining, it was confirmed as Gram-negative, occurring in short rod shape. The bacteria test reagent strips of Merieux Alliance ID32 E was selected for the biochemical identification.

2.5 PCR identification of isolated strain

Primer design: the most significant 16S rRNA in bacterial taxonomy was taken as the target gene and th e primer was composed by Shanghai Invitrogen Life Technologies Co., Ltd. The sequence of the primer was shown as follows:

Seq forward: 5'-AGAGTTTGATCCTGGCTCAG-3'

Seq reverse: 5'-GGCTACCTTGTTACGACTT-3'

The special primer is designed according to the aerolysin gene sequence of the Streptococcus dysgalactiae. The target fragment is 597bp[5]. The primer sequence is shown as follows:

Seq forward: 5'-GAGAAGAAGCCCAGAGCG-3'

Seq reverse: 5'-AGTTGGTGGCAGTATCGTAA-3'

PCR response: DNA was extracted with boiling method, Taq DNA polymeras, dNTPs, DL2000 molecular weight standard was purchased from Takara Company. PCR reaction condition: 1.0 μL of 50 μL reaction system, 10×PCR-Buffer 5.0 μL, 25 mmoL/L MgCl2 4.0 μL, 2.5 mmoL/L dNTP 4.0 μL, 5 U/μL *Taq* polymerase 0.3 μL, 10 pmoL/L upstream and downstream primer shall be added into each tube successively, and each tube shall be added with ultrapure water 33.7 μL and template 1 μL successively. After initial denaturation at 94℃ for 3 min, the PCR would be amplified in the following conditions: 94℃ 1min, 54℃ 1min, 72℃ 1min, 30 cycles, 72℃ extension for 7 min. PCR product shall be tested with 1.5% agarose gel electrophoresis, and ethidium bromide shall be employed for dyeing [5]. The PCR product sequencing is accomplished by Shanghai Invitrogen Life Technologies Co., Ltd. and afterwards, Blast analysis was conducted.

1.6 Drug susceptibility test

The drug allergy paper agar plate diffusion method was purchased from Hangzhou Binhe Microorganism Reagent Co., Ltd. 15 μL of bacteria in shake culture was covered on the drug allergy paper agar plate evenly, according to the antibacterial drug sensitivity test standard (SAST) [6].

3. Result

3.1 Microscopic observation of the mycelia morphology

The Gram staining was negative in short rod shape. The bacterial colony was smooth, dimpling, rounding and in flesh color on the regular broth plate.

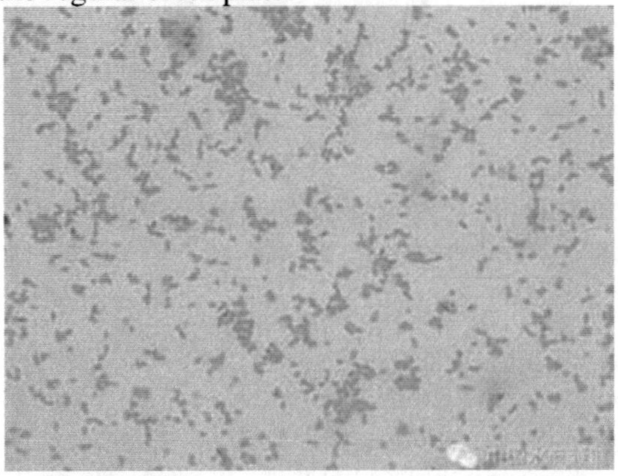

Figure2 Gram stain of separation bacteria

3.2 Physiological and biochemical identification

The single colony was cultivated on plain agar plate at 30°C for 24 hours, and the sub-cultivated BioMerieux ID32E enteric bacilli identification stripe was cultivated at 37°C for 24 hours. The result in biochemical identification tube of bacterial has shown as follows:

Tab. 1 The identification results of Strain by BioMerieux ID32E Identification System

Test items Result		Test items Result	
Oxidase	+	Catalase	+
Power	+	Glucose	+
Lactose	-	arginine hydrolase	+
lysine decarboxylase	-	Urea	-
ornithine decarboxylase	-	Esculin	+
Gelatin	+	Arabinose	+
Mannose	+	Mannitol	+
Pyruvate	+	salicylic acid	+
salt-free peptone water	+	6% saline	-
3% saline	+	10% saline	-

HEC toxin was detected with blood agar plate, and it turned out that there was clear hemolysis cycle around the bacterial colony, which was positive. The isolate was characterized by typical biochemical characters of aeromonas hydrophila[7] .

3.3 PCR identification of isolated strain

The 16S rRNA gene segment from PCR amplification was 1500 bp, and recycling sequencing was conducted for the segment. It could be discovered through the homology analysis of sequence that it had 99% homology with the conservative property of the standard strain ATCC7699 of aeromonas hydrophila.

The special primer has a clear and bright belt at 597bp (figure 2). The belt 1 is for golden staphylococcus ATCC29213.The belt 2 is for blank contrast, the belt 3 is for the separated bacterial strain, the belt M is for Marker.

Combining the morphology and physiological-biochemical characteristics, the isolate strain was identified as aeromonas hydrophila[8].

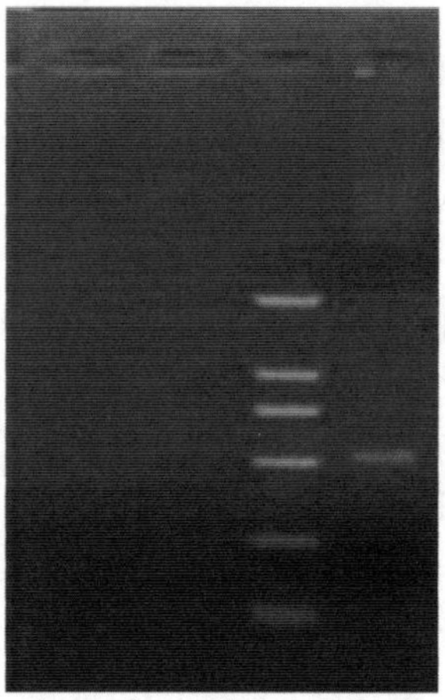

Figure 2 PCR identification of special aerolysin gene of separate strain

3.4 Drug susceptibility test

It turned out that cefradine, lincomycin, norfloxacin, norfloxacin, cephalexin, penicillin, streptomycin, kanamycin and furazolidone could not restrain the bacterial, but the isolate was sensitive to the doxycycline, chloramphenicol, enrofloxacin, Ciprofloxacin, etc. The diameter of inhibition zone /mm of several antibacterial agents to the bacterial strain is shown in table 1.

Table 2 the diameter of inhibition zone /mm of several antibacterial agents to the bacterial strain

Antibiotics	Diameter of inhibition zone /mm	Result	Antibiotics	Diameter of inhibition zone /mm	Result
Tobramycin	16	S	Chloramphenicol	24	S
Cefradine	10	M	Enrofloxacin	30	S
Gentamicin	14	S	Ciprofloxacin	34	S
Macrodantin	16	S	Spectinomycin	25	S
Lincomycin	0	R	Kanamycin	8	R
Norfloxacin	10	R	Rocephin	33	S
Cephalexin	0	R	Neomycin	14	S
Doxycycline	18	S	Furazolidone	8	R
Penicillin	0	R	Streptomycin	6	R

Notes: S. sensitive; R. drug resistance; M. mediation

3.5 Clinical symptoms and artificial infection test of procambarus clarkii

The main symptoms of infected procambarus clarkii were: at the first beginning, there was no clinical manifestation, and when it ws attacked by the disease, the pathogeny of large-scale death of procambarus clarkia was faint and slow in action, it could be seen from the dissection that the hepatopancreas was swelling, the intestinal tract was filled with blue blood and without food. On the third of the attack, the procambarus clarkii started to die. In later period, the intestinal tract of procambarus clarkii was congestive. The entire test lasted for eight days[9].

Table 3 artificial infection test result

Concentration of bacterium /cfu	Dosage /mL	Procambarus clarkia for testing	Mortality
1.0×108	0.1	7	7
1.0×107	0.1	7	6
1.0×106	0.1	7	4
1.0×105	0.1	7	3
1.0×104	0.1	7	1
Normal saline	0.1	7	1
Blank control	0.1	7	0

It can be seen from the challenge attack that this strain had strong toxicity for healthy procambarus clarkii, and LD50 reached 5.3×105cfu/mL. The intestinal tract and pancreatic tissue of procambarus clarkii dying on the fourth day of challenge test was selected for the isolation of bacteria, and through formation and biochemical identification proof, the bacterial strain with consistent growth characteristics of strain N201307 could be obtained. Meanwhile, the bacterial colony on the isolation plate was pure and free from infectious microbe. It failed to isolate the bacterial strain from the dead procambarus clarkii in normal saline group.

4. Discussion

Aleromonas hydrophila drew the attention of medical workers since it could result in diarrhea, and later, it was widely discovered in fresh water, sludge, soil and excrement. It consisted of pathogenic strains and non- pathogenic strain. The former was pathogenic for aquatic animals, beasts and birds, and human, as a kind of typical pathogenic bacteria resulting in the disease among human, beat and fish [10]. Aeromonas was dangerous in the aquaculture, and it may give rise to several kinds of bacterial blood poisoning, such as grass carp, mandarin fish, Tilapia mossambica, carp, crucian, sturgeon, etc. [11] Procambarus clarkii could be discovered in natural water body, such as lakes, rivers, ponds, ditch, rice field, etc. distributing in dozens of provinces and regions, including Jiangsu, Zhejiang, Hubei, Hunan, Anhui, Shanghai, Shandong, etc. of China [12]. Previously, the procambarus clarkii was captured from natural resources, and in recent years, the enthusiasm of culturing procambarus clarkii in aquaculture industry is on the rise. According to the statistics, the water area for the culture of procambarus clarkii in Anji of Huzhou has already reached 400 hectares, close to millions of Yuan. As the emerging cultivation variety, there are still few studies on the disease of procambarus clarkii at home and abroad, and the attention is insufficient. The current domestic and foreign research results have shown that the main pathogenic microorganism resulting in the substantial death of procambarus clarkii are WSSV [13] and Vibrio Parahaemolyticus) [14]. For this disease of procambarus clarkii in Anji, pathogenic microorganism is isolated from the focus and it is identified as aeromonas hydrophila. It is tested from the challenge test that LD50 is 5.3×105cfu, and according to the standard for the division of toxicity of aeromonas hydrophila by Mittal[15] et al. LD50 was of strong toxicity when it ranges between 104 and 105 cfu, weak toxicity between 106 and 107 cfu, no toxicity when it is larger than 108 cfu. The isolate N201307 has strong toxicity, which is in accordance with the substantial death occurred in the culture zone of procambarus clarkii in Anji County.

It is reflected by the test result of the antibacterial effect of different antibiotics on the isolate N201307 that cefradine, lincomycin, norfloxacin, cephalexin, penicillin, streptomycin, kanamycin, and furazolidone have strong tolerance to drugs, and it may be related to the category of bacterial isolates and frequency of applying the antibiotics in the culture process [16]. The isolate is sensitive to several kinds of antibiotics, such as the doxycycline, chloramphenicol, norfloxacin, Ciprofloxacin, etc. but among them, several are forbidden in the aquaculture, which cannot be applied even if they have good effects. Besides, it also shows that in the current high-density culture model, the procambarus clarkii which is supposed to grow well in severe natural environment is experiencing the declining resistivity to the pathogenic microorganism. Besides, it shall pay attention to the prevention of disease in the culture process.

ACKNOWLEDGEMENT

The authors gratefully acknowledge the financial supports by Science Technology Department of Huzhou City (Project number: 2013C102070).

References

[1] Chen Changfu, Yang Jun, Liu Yuangao, et al, Pathogenic Organisms for Explosive Epidemic Disease of Red Swamp Crayfish, Procambarus Clarkii [J]. Journal of Huazhong Agricultural University, 2008, 27 (6):763-67.

[2] Lei Zhiwen, Huang Jie, Shi Chengyin, et al. Investigation into the Hosts of White Spot Syndrome Virus (Wssv) [J]. Oceanologia Et Limnologia Sinica, 2002, 33(3): 250-258.

[3] Zhou Yuhua, Cai Ke, Isolation and Identificaiton of Soft-Shelled Turtle Aeromonas Hydrophila [J]. Inspection and Quarantine Science, 2002, 12 (3): 48-49.

[4] JIMENEZ R, ROMERO X. Infection by intracellular bacteriumin red claw crayfish, Cherax quadricarinatus (Von Martens) ,in Ecuador [J] .Aquaculture Research, 1997, 28 (12): 923-929.

[5] Chu Weihua , Lu Chengping. PCR detection of pathogenic Aeromonas hydrophila by specific 16S rDNA and aerolysin gene[J]. Journal of fishries of china, 2005, 29(1):79-82.

[6] Wang Guoliang, Xu Yijun, Jin Shan, Identification and phylogenetic analyses of Nocardia in snakehead, Ophiocephalus argus Cantor [J]. Journal of Fisheries of China, 2008, 2 (3): 449-454.

[7] Azad I S, Shankar K M , Mohan C V, et al. Biofilm vac2cine of *Aeromonas hydrophila* standardization of doseand duration for oral vaccination of earos [J]. Fish & Shellfish Immunology, 1999(9): 519-528.

[8] Sambrook J, Russell D W.Molecular cloning a laboratory manual [M].Beijing: Science Press.2002, 387-392.

[9]. Tong Zhaowei, Zhang Longqi, Wang Weihong, et al. Analysis the infection and antimicrobial resistance of Aeromonas hydrophila [J]. Chinese Journal of Microecology, 2008, 20 (1): 75-76.

[10] Saitanu K, Wongsawang S. Red-score disease in carp (Cyprinus carpio L.) [J]. J Aquat Anim Health, 1982, 5(3): 79-86.

[11] Costerton j w, Stewart P S, Greenberg E P, et al. Bacterial biofilm: A common cause of persistent infections [J]. Science, 1999, (284): 1318-1322.

[12] Jha R K, Xu Zirong, Bai Shijuan, et al. Protection of *Procambarus clarkii* againstwhite spot syndrome virus using recombinant oral vaccineexp ressed in *Pichia pastoris* [J]. Fish & Shellfish Immunology, 2007, 22: 295-307.

[13]. Chen Changf, Liu Yuangao, He Guangwen, et al. The Bacterial Pathogen of Outbreak Disease in Procambarus clarkii [J]. Journal of Huazhong Agricultural University, 2009, 28 (3): 193-197.

[14] Hong Xupeng, Lu Hongda, Zhang Qinghua, et al. Effect of Astragalus Polysacchairide on the Anti-Infection of White Spot Syndrome Virus (WSSV) in Procambarus Clarkia [J]. Journal of Shanghai Ocean University, 2014, 23(3):423-428.

[15] Jiravanichpaisal P, Roos S. A hightly virulent pathogen, Aeromonas hydrophila, from the freshwater crayfish Pacifastacus leniusculus [J]. J Invertebr Pathogen, 2009, 101(1): 56-66.

[16] Mittal K R, Lalonde G, Leblanc D, et a1.Aeromonas hydrophila in rainbow trout: relation between virulence and surface characterics [J]. Can J Microbiol, 1980, 26: 1501-1503.

[17]. Chen Honglian, Song Guangtong, He Jixiang, et al. Isolation, Identification and Antibiotics Susceptibility Test of Citrobacter Freundii from Procambarus Clarkii [J]. Freshwater Fishery, 2014, 44 (1): 73-77.

Ecological values and costs of urban green space under different management modes

Chengxiang Zhang, Jinhua Sun*, Zhe Zhang, Wei Wang
Chinese Research Academy of Environmental Sciences, Beijing 100012, China

Abstract: On the basis of ecological service function and negative influences on environment of urban green space analysis, idea about eco-management and agricultural management has been put forward. An urban green space ecological value, cost and profit evaluation model was established by virtue of previous research about accounting of ecology value. Then green space of road was taken as a case to evaluate its ecological values and costs under different management modes. The results showed that irrigation water resources consumption amount is greater than the green water conservation quantity universally existing in the management of urban green space. In the general management of urban green space, use of chemical fertilizer and pesticide brought great environmental pollution. In the eco-management mode, profit of urban green space was nearly 50% higher than that in the general mode, for cost of urban green space reduced. Urban green space has great economic output value while its ecological cost almost the same as that in general management mode. Through this study we suggest that urban green space should be managed either in the eco-management mode or in the agricultural mode, so that human been could get more ecological benefits.

Keywords: urban green space; ecological value; ecological cost; management mode

1 Introduction:

With the rapid development of urbanization, urban green space is drawing more and more attention due to their ecological function such as air cleaning, living environment perfection (Costanza R,d' et al. 1997). As an ecosystem approaching to human living space, urban green space and its ecological benefits are directly related to the quality of human life, and are significantly affected by human management including fertilization, pest control, irrigation, pruning, etc. At present, people maintain and manage urban green space according to the National relevant technical regulation in general modes, such as utility of chemical fertilizers, pesticide in controlling pest and so on. However, ecological ideas such as organic manure instead of chemical fertilizers, physical and biological means of pest control are seldom used in the urban green space management. While urban green space supplied ecological service function and ecological benefits, it also caused environmental pollution and water resource consumption (named as the ecological costs in this article), such as water and air pollution caused by fertilizer and pesticide use, carbon emissions and air pollution mechanized pruning shrubs and lawns sheet caused by the consumption of water for irrigation of urban green space. Water and air pollution caused by fertilizer and pesticide use during the urban green management, carbon emissions and air-pollution from mechanized pruning shrubs and lawns sheet, and consumption of water for irrigation of urban green space. Since Costanza, etc. carried out quantitative evaluation to global ecosystem services in 1997 and Nature's Service: Societal Dependence on Natural Ecosystems edited chiefly by Daily having be published, the value of ecosystem services has become a hot and the frontier research

* Corresponding author. sunjinhua2004@126.com

area in Ecology, Ecological Economics, etc(Costanza R,d' et al. 1997; Daily G C 1997). For now, the value of urban green space researches is mainly focused on landscape service value(Pang J 2008; Zhou S L 2007;Sun M et al. 2010). However, there were only a few studies on the ecological value indicating simple estimate of some ecological value and Comparison of adaptability among different green vegetation (Tyrvainen, L 2001; Manfred K 2004). The studies of resource consumption and environmental pollution are rarely concerned, and comparative study between urban green ecological value and ecological costs is basically blank in the home and aboard(Wang W W 2011; Jari L and Maijia S 2009).

Green space of road is an important part of the urban green space. Taking the urban road greening project as a case, this paper aims to estimate the value of ecosystem services and ecological costs, and propose the idea of agricultural management, i.e.to use local fruit trees instead of green trees, reasonable grazing instead of lawns and shrubs trimmed etc, and compare the ecological value and cost generated from different urban management modes, including general management mode, eco-management mode and agricultural management mode, for further exploring reasonable and scientific management mode for urban green, so as to gain greater ecological service function and ecological value.

2 Research methods

Ecological value is— the service value that ecosystem could provide for humanity. An ecological system has 23 kinds of ecology functions in 4 general categories including production function, regulatory function, biological diversity maintenance function, information function. For urban green space, major functions include gas regulation, environment purification, water conservation and noise reduction. Compared with general management and eco-management, it should be noted that urban green space has economic output function in the mode of agricultural management, while its ecological cost is the same with general management. In the eco-management mode, urban green space value was the same as that in the general management mode, while its ecological cost caused by fertilizer could be neglected, and its pollution from pesticide cost was just one half of that in the general management mode. In this study we define ecological profit as the difference between ecological value and ecological cost. Ecological value research method is the evaluation of ecological value by using ecological economics theories, such as market value method, production cost method, substitute expense method, and so on.

2.1 Value and cost of gas regulation service

2.1.1 Value and cost of Carbon sequestration

The methods of evaluating value and cost of carbon sequestration include carbon tax method (belong to market value method) and reforestation cost method (belong to production cost method). In this study, carbon tax method was chose and carbon tax in Sweden was used for its widespread recognition..

Formula for calculating carbon sequestration value is: $V_C=S\times Q_C\times T_C$. In this formula, Vc means carbon sequestration value of urban green space (yuan RMB·a^{-1}), S means the area of urban green space (m^2), Q_C means annual carbon sequestration capability per unit of urban green space area($t\cdot m^{-2}\cdot a^{-1}$), T_C means carbon tax in Sweden(150 \$·t^{-1}C; Anderson D 1990), the exchange rate between the US dollar to RMB yuan is about 1:6.2.

Carbon emission is from mechanical pruning of lawn and shrubbery, and N_2O volatilized from fertilizer can be converted to carbon according GWP, so formula of calculating carbon sequestration cost is:

$$V_C = \sum_{i=1}^{n}(\frac{M_i \times S_i}{V_i} \times W_i \times Q_{Ci}) \times T_C + \sum_{i=1}^{n}(N_i \times S_i) \times Q_{N_2O} \times K \times T_C$$

In this formula, V_c means carbon sequestration cost of urban green space(yuan RMB·a^{-1}), M_i means frequency of pruning urban green vegetation(tines·a^{-1}), S_i means area of urban green vegetation(m^2), V_i means prune speed($m^2 \cdot h^{-1}$), W_i means power of the prune machinery(kw), Q_{Ci} means carbon emission intensity of the prune machinery(g·kwh^{-1}), N_i means annual amount of N fertilizer application per unit of urban green space area(t·$a^{-1} \cdot m^{-2}$), Q_{N2O} means loss rate of N fertilizer in the form of N_2O(0.67%; Lai L et al. 2009), K means conversion rate between N_2O to carbon according GWP.

2.1.2 Value and cost of Releasing oxygen

Shadow engineering approach was used to evaluate value and cost of releasing oxygen of urban green space. Formula of calculating oxygen release value is: $V_{O2}=S \times Q_{O2} \times P_{O2}$. In this formula, V_{O2} means releasing oxygen value of urban green space(yuan RMB·a^{-1}), S means the area of urban green space(m^2), Q_{O2} means annual oxygen release capability per unit of urban green space area(t·$m^{-2} \cdot a^{-1}$), P_{O2} means cost of industrial process of oxygen making(700 yuan RMB·t^{-1}; Zhang X L et al. 2011).

Oxygen consumption is from mechanical pruning of lawn and shrubbery, formula is:

$$V_{O2} = \sum_{i=1}^{n}(\frac{M_i \times S_i}{V_i} \times W_i \times Q_{O2i}) \times P_{O2}$$

In this formula, V_{O2} means oxygen consumption cost of urban green space(yuan RMB·a^{-1}), M_i means frequency of prune urban green vegetation(tines·a^{-1}), S_i means area of urban green vegetation(m^2), V_i means prune speed($m^2 \cdot h^{-1}$), W_i means power of the prune machinery(kw), Q_{O2i} means oxygen consumption intensity of the prune machinery(g·kwh^{-1}), P_{O2} means cost of industrial process of oxygen making(700 yuan RMB·t^{-1}).

2.1.3 Value and cost of Humidity increase

Plants can increase humidity of air through transpiration, and no special investment is needed in this process, so humidity increase cost is neglected in this study. Equivalent substitution approach is used to evaluate humidity increase value, formula is: $V_{Tr}=S \times P_{Tr}$. In this formula, V_{Tr} means humidity increase value of urban green space(yuan RMB·a^{-1}), S means area of urban green space(m^2), P_{Tr} means humidity increase value per unit of urban green space area(104.2 yuan RMB·m^{-2}; Tang X H 2011).

2.1.4 Value and cost of Air temperature reduction

Plants can reduce air temperature through two ways: capturing solar energy through photosynthesis and gasifying water through transpiration. In these processes no special investment is needed, so air temperature reduction cost is neglected in this study. Formula of calculating air temperature reduction value is: $V_T = S \times (Q_C \times W_P + E_{Tr}) \times P_T$.

In this formula, V_T means air temperature reduction value(yuan RMB·a^{-1}), S means area of urban green space(m^2), Q_C means annual carbon sequestration capability per unit of urban green space area in summer(t·m^{-2}·a^{-1}), W_P means solar radiation absorptive intensity(39.1kJ·g^{-1}), E_{Tr} means annual absorb caloric per unit of urban green space area through transpiration in summer(kJ·m^{-2}·a^{-1}), P_T means cost of air temperature reduction by air-conditioning(One air-conditioning can 5225 kilojoules, at the same time it will expends 0.82 kWh electricity, that means about 0.41 yuan RMB; Wang C J 2008).

2.2 Value and cost of Environment purification

2.2.1 Environment purification value

Substitute expense method was used to evaluate the value of environment purification of urban green space. Environmental governance cost is used to substitutes environment purification value. Because of fertilizer and pesticide application, only the parts of fertilizer and pesticide which have not been absorbed by plants can lead to environment pollution, so water and soil purification values are neglected in this study.

（1）SO_2 absorption value

Formula is: $V_{SO2}=S\times Q_{SO2}\times F_{tSO2}$. In this formula, V_{SO2} means SO_2 absorption value(yuan RMB·a^{-1}), S means area of urban green space(m^2), Q_{SO2} means annual SO_2 uptake per unit of urban green space area(t·m^{-2}·a^{-1}), F_{tSO2} means cost of industrial process of SO_2 uptake(3000 yuan RMB·t^{-1}; Zhang X L et al. 2011).

（2）NO_x absorption value

Formula is: $V_{NOx}=S\times Q_{NOx}\times F_{tNOx}$. In this formula, V_{NOx} means NO_x absorption value(yuan RMB·a^{-1}), S means area of urban green space(m^2), Q_{NOx} means annual NO_x uptake per unit of urban green space area(t·m^{-2}·a^{-1}), F_{tNOx} means cost of prevention of automobile exhaust NO_x pollution(16000 yuan RMB·t^{-1}; Zhang X L et al. 2011).

（3）Dust removal value

Formula of calculating dust removal value is: $V_D=S\times Q_D\times F_t$. In this formula, V_D means dust removal value(yuan RMB·a^{-1}), S means area of urban green space(m^2), Q_D means annual effects of dust removal per unit of urban green space area(t·m^{-2}·a^{-1}), F_t means cost of industrial process of dust uptake(170 yuan RMB·t^{-1}; Zhang X L et al. 2011).

2.2.2 Environment pollution cost

Urban green space causes Environment pollution cost via three ways: NO_x emission from mechanical pruning of lawn and shrubbery, NO_x and NH_3 volatilized from nitrogen fertilizer, soil and water polluted by fertilizer and pesticide。

（1）Cost of NO_x emission from mechanically pruning of lawn and shrubbery

Formula is: $V_{NOx}=\sum_{i=1}^{n}(\frac{M_i\times S_i}{V_i}\times W_i\times Q_{NOxi})\times F_{tNOx}$。In this formula，$V_{NOx}$ means NO_x emission cost(yuan RMB·a^{-1}), M_i means frequency of prune urban green vegetation(tines·a^{-1}), S_i means area of urban green vegetation(m^2), V_i means prune speed(m^2·h^{-1}), W_i means power of the prune machinery(kw), Q_{NOxi} means NO_x emission intensity of the prune machinery(g·kwh^{-1}), F_{tNOx} means cost of prevention of automobile exhaust NO_x pollution(16000 yuan RMB·t^{-1}).

(2) Cost of NO_x and NH_3 volatilization of Nitrogen fertilizer

Formula is: $V_N = \sum_{i=1}^{n}(N_i \times S_i) \times Q_{NH_3} \times T_{NH_3} + \sum_{i=1}^{n}(N_i \times S_i) \times Q_{NO_x} \times F_{tNO_x}$. In this formula, V_N means NO_x and NH_3 volatilization of Nitrogen fertilizer cost(yuan RMB·a^{-1}), N_i means annual amount of N fertilizer application per unit of urban green space area or per plant(t·a^{-1}·m^{-2} or t·a^{-1}·$plant^{-1}$), S_i means area of urban green vegetation or plants(m^2 or plant), Q_{NH3} means loss rate of N fertilizer in the form of NH_3(5%; Lai L et al. 2009), T_{NH3} means cost of NH_3 pollution purification by air purifier(361455.7 yuan RMB·t^{-1}), Q_{NOx} means loss rate of N fertilizer in the form of NOx(0.5%; Lai L et al. 2009), F_{tNOx} means cost of prevention of automobile exhaust NO_x pollution(16000 yuan RMB·t^{-1}).

(3) Water pollution from fertilizer cost

Formula is: $V_{FW} = \sum_{i=1}^{n}(N_i \times S_i) \times Q_{NW} \times T_{NW} + \sum_{i=1}^{n}(P_i \times S_i) \times Q_{PW} \times T_{PW}$. In this formula, V_{FW} means water pollution from fertilizer cost (yuan RMB·a^{-1}), N_i means annual amount of N fertilizer application per unit of urban green space area or per plant(t·a^{-1}·m^{-2} or t·a^{-1}·$plant^{-1}$), S_i means area of urban green vegetation or plants(m^2 or plant), Q_{NW} means loss rate of N fertilizer into water(10%), T_{NW} means cost of TN uptake in sewage treatment plants(5000 yuan RMB·t^{-1}; Chinese Research Academy of Environmental Sciences 2012), P_i means annual amount of P fertilizer application per unit of urban green space area(t·a^{-1}·m^{-2}), Q_{PW} means loss rate of P fertilizer into water(10%), T_{PW} means cost of TP uptake in sewage treatment plants(5000 yuan RMB·t^{-1}; Chinese Research Academy of Environmental Sciences 2012).

(4) Cost of soil pollution via fertilizer

Formula is: $V_{CF} = \sum_{i=1}^{n}(N_i \times S_i) \times P_{CF}$. In this formula, V_{CF} means soil pollution from fertilizer cost(yuan RMB·a^{-1}), N_i means annual amount of fertilizer application per unit of urban green space area or per plant(t·a^{-1}·m^{-2} or t·a^{-1}·$plant^{-1}$), S_i means area of urban green vegetation or plants(m^2 or plant), P_{CF} means cost of soil pollution from per unit of fertilizer(300 yuan RMB·t^{-1}).

(5) Pollution from pesticide cost

On the basis of previous research, pollution from pesticide cost is considered to be 1% of crop farming output value(Chen D D et al. 2008). Formula is:

$V_P = V_G \times 0.01 \times \frac{S_U}{S_A + S_U}$. In this formula, V_P means pollution from pesticide cost (yuan RMB·a^{-1}), V_G means crop farming output value(yuan RMB·a^{-1}), S_U means area of urban green vegetation(m^2), S_A means land area used for planting(m^2).

2.3 Value and cost of Water conservation

Shadow engineering approach used to evaluate water conservation value and cost of urban green space. Formula of calculating water conservation value is: $V_{H2O}=S \times Q_{H2O}/1000 \times K \times P_{H2O}$. In this formula, V_{H2O} means water conservation value(yuan RMB·a^{-1}), S means area of urban green vegetation(m^2), Q_{H2O} means mean annual precipitation(mm), K means water holding rates(25%), P_{H20}

means cost of building reservoir(0.67 yuan RMB·m^{-3}; Ouyang Z Y et al. 1999).

Water conservation cost is from irrigation, formula is:

$$V_{H_20}=\sum_{i=1}^{n}(N_i \times S_i \times Q_{H_20i})\times P_{H_2O}\cdot$$

In this formula, V_{H2O} means water conservation cost(yuan RMB·a^{-1}), N_i means frequency of irrigate urban green vegetation(tines·a^{-1}), S_i means area of urban green vegetation or plants(m^2 or plant), Q_{H2Oi} means irrigation water amount per unit of urban green space area(m^3·m^{-2} or m^3·$plant^{-1}$), P_{H20} means cost of building reservoir(0.67 yuan RMB·m^{-3}).

2.4 Value and cost of Noise reduction

Reforestation cost approach was used to evaluate value and cost of noise reduction. Formula of calculating noise reduction value is: $V_N= S\times F\times C\times 0.15$. In this formula, V_N means noise reduction value(yuan RMB·a^{-1}), S means the area of urban green space(m^2), F means mean reforestation cost(yuan RMB·m^{-3}), C means timber yield per unit of forest area(m^3·m^{-2}·a^{-1}).

Cost of noise reduction is caused by mechanically pruning of lawn and shrubbery, noise reduction cost is neglected in this study for it's an occasional work.

2.5 Economic output value and cost

Economic output aims at agricultural management, compared with general management, no more cost is needed , so economic output cost is neglected in this study(Primmer E and Furman E 2012). Formula of calculating economic output value is: $V_A=N_F\times C_F\times P_F+N_L\times S\times C_L\times P_L$. In this formula, V_A means economic output value, N_F means quantities of trees that replaced by fruit trees（plant），C_F means yield of fruit(kg·$plant^{-1}$·a^{-1}), P_F means average price of the fruit（yuan RMB·kg^{-1}），N_L means percentage of the area of urban green space that replaced by grazing，S means the area of urban green space（m^2），C_L means grazing capacity（cow·m^{-2}·a^{-1}），P_L means average price of animal（yuan RMB·cow^{-1}).

3. Case study

The urban road greening engineering of Yongjia county(Wenzhou city, Zhejiang province) in 2012 was used in the case study(Table 1). Ecological value, ecological cost and profit were analyzed in the condition of general management(actual), eco-management(hypothesis) and agricultural management mode(hypothesis) . Yongjia county is located in the southeast coastal areas of China, where climate type is belong to subtropical monsoon climate and the annual average rainfall is 1718.3 millimeters. The urban greening level of Yongjia is exceeded the average level of China. The integrated analysis showed that greening engineering construction and management of Yongjia is significantly representative.

Table1 Vegetation statistical of Yongjia urban road greening engineering

Vegetation type	Extra-large tree(plant)	General tree (plant)	Shrub space (m^2)	Lawn area (m^2)
Number	4204	18081	62775	17363

3.1 Ecological value and cost calculation of urban green space

Considering trees, lawns and shrubs were usually mixed cultivation in the urban greening, the sum of lawn areas and brush areas was used as the urban green space areas in this study. Table2 shows the parameters for the ecological value and cost calculation of urban green space.

Table2 Parameters for the ecological value and cost calculation of urban green space

Parameter	Value	Parameter	Value	Parameter	Value
Carbon sequestration	$4.36t \cdot hm^{-2} \cdot a^{-1}$	Releasing oxygen	$11.80t \cdot hm^{-2} \cdot a^{-1}$(Zhou B B and Li Z K 2000)	SO_2 absorption	$88.65kg \cdot hm^{-2} \cdot a^{-1}$
NOx absorption	$0.38t \cdot hm^{-2} \cdot a^{-1}$(Peng J 2005)	Amount of chemical fertilizer	general management level	Agricultural production of Yongjia county 2013	465.4 acres of land is used for planting, output value is 200 million yuan RMB
Irrigation	general management level	Reforest cost	240.03 yuan $RMB \cdot m^{-3}$	Timber yield per unit of forest area	$80\ m^3 \cdot hm^{-2}$
Fruit used to replace trees	*Myrica rubra*(assume that 30% general trees are replaced; Agriculture Bureau of Yongjia 2011)	Animal used in the mode of agricultural management	cow		

3.2 Ecological profit at different management modes

Compared the ecological profit in different management modes, we concluded that the profit of the three modes is: eco-management mode > agricultural management mode > general management mode(Figure1). There is the same ecological value between eco-management mode and general management mode. Due to none fertilizers and pesticides used, which reduced the ecological cost in the mode of eco-management, the ecological profit is nearly 50% higher than general management mode. The ecological profit margin is four times higher than general management. Because vegetable plantation and graze can make profits, ecological profit in the mode of agricultural management mode is 16.6% higher than general management. So there is a big potential in improving urban green space ecological value via better use of urban green space(such as agriculture production).

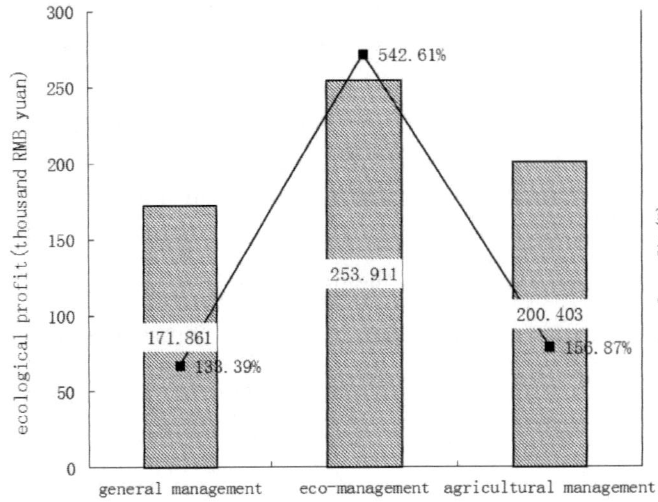

Figure1 Ecological profits of urban green space at different management mode

As shown in Table3, environment purification profits of urban green space in the mode of general management and agricultural management are all negative, it could be indicated that the pollution caused by pesticides and fertilizers offset the environment purification profits brought by green space itself. This may be the reason that the improved environment quality is limited in recent years with the increased urban greening. The water consumption in the management of urban green space is two times of the water amount of conserved by urban green space itself, which induced the conserved water profit in the urban green space is negative value. Urban green space in the mode of general management and eco-management could not produce economic value, but the economic value in the mode of agricultural management can reached nearly 20% of total ecological profits of general management, so the economic profits is considerable if apply agricultural management in the management of urban green space.

Table3 Major ecological function profits analysis of urban green space

Management mode	Gas regulation			Environment purification			Water conservation			Noise reduction			Economic output			Total		
	Cost	Value	Profit	Cost	Value	Profit	Cost	Value	Profit	Cost	Value	Profit	Cost	Value	Profit	Cost	Value	Profit
General management	4907	189899	184992	79396	64703	-14693	44541	23065	-21476	0	23038	23038	0	0	0	128844	300705	171861
Eco-management	671	189899	189228	1582	64703	63121	44541	23065	-21476	0	23038	23038	0	0	0	46794	300705	253911
Agricultural management	4236	189899	185663	78973	64703	-14270	44541	23065	-21476	0	23038	23038	0	27448	27448	127750	328153	200403

3.3 Ecological cost analysis of general management mode

The ecological cost of urban green space in the mode of general management is very big, the ecological cost of atmospheric contamination caused by fertilizer account for nearly 60% of total ecological cost, and the ecological cost caused by irrigation water account for almost 35% of total ecological cost. It is an urgent problem to solve the low efficiency of fertilizer and water resource utilization. There are several limitations in our research, including lack of basic research and quantified data. We only calculated the direct economic loss of agriculture production caused by pesticides, and the results shown that the ecological cost of pesticides utilization in urban green space is extremely low. This results were not assistant with the actual situation that pesticides utilization damaged the ecological environment. In fact, pesticides induced a big damage to ecological environment, for example, pesticides damage drinking water, human health and beneficial organisms. This research field needs to be more strength in the future.

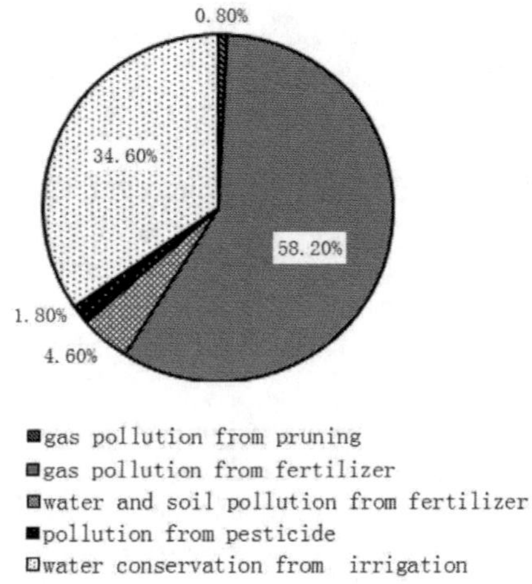

Figure2 Ecological cost analysis of urban green space in the mode of general management

Conclusion and discussion

The study demonstrated that the use of chemical fertilizer and pesticide brought great environmental pollution in the general management of urban green space, which offset the value of clean environment green itself to some extent, and the profit of ecological environment purification is negative. If the result of environmental pollution and ecological destruction from chemical fertilizer and pesticide production or transport is considered, the pollution is virtually transferred from town to the rural or other remote areas where fertilizer and pesticide factory is located in, during the urban green space management providing ecological services(Fare R and Grosskop S L 2006). Irrigation water resources consumption amount is greater than the green water conservation quantity universally existing in the management of urban green space, so the drought is probably shifted from town to the nearby water sources in arid during the urban green water generating the value of water conservation.

Compared with the general urban green management mode, if the eco-management mode was taken, the ecological cost of urban green space ecological can be reduced, thus the ecological profit of environment purification of urban green can be increased. Agricultural management mode can brought some economic profit by the way but do not increase ecological cost. On the one hand it is suggested to promote the eco-management mode, which was composed of adopting water-saving irrigation, soil testing and fertilizer, organic fertilizer, low toxicity or biological pesticides and physical pest control measures, reducing the amount of chemical fertilizer and pesticide, so as to decrease the ecological cost of urban green space and promote transfer urban green management mode from landscape oriented to ecological oriented. On the other hand, the implementation of agricultural management can perhaps transfer the maintenance and management of urban green from the behavior of the government fiscal investment purely to the behavior of production and management to enhance the economic sustainability of urban green space management, if the condition permits.

The model parameters are mainly referred to the existing domestic and foreign achievements in this study. There is a need to carry out the calibration of model parameters to improve the ecological value and the cost of computing in the future research because of the lack of consideration of regional differences. Due to the weak fundamental researches related to ecological value and costs, so some effects of urban green space on the ecological environment cannot be quantified through indicators, such as possible ecological and environmental risk from eco-management, the natural sedimentation after fertilizer volatilized into the air, continuous damage of pesticide on the ecological system, which are the key research hotspots to be strengthened in the future.

References

[1] Agriculture Bureau of Yongjia(2011).Yongjia modern agriculture function plans.
[2] Anderson D(1990). Carbon emissions and carbon fixing from an economic perspective// Forestry Economics Research Conference . Canada ,York University: 26-54.
[3] Chen D D, Gao W S, Sui P, Wu T L(2008). Dynamic analysis on energy efficiency of modern planting system and grain production—a case study of Luancheng, Hebei. Progress in Geography, 27(1): 99-104.
[4] Chinese Research Academy of Environmental Sciences(2012). Investigation and Evaluation about Ecological Safety of Lakes. Beijing: Science Press.
[5] Costanza R,d' Agre R, de Groot R, Farber S, Grasso M, Hannon B, Limburg K, Naeem S, O'Neill R V, Paruelo J, Raskin R G, Sutton P, vanden Belt M(1997). The value of ecosystem service and natural capital in the word. Nature. 387(6630): 253-260.
[6] Daily G C(1997). Nature's service: societal dependence on natural ecosystems. Washington D C: Island Press: 1-78.
[7] Fare R, Grosskop S L. Weber W(2006). Shadow price and pollution costs in U. S agriculture. Ecological Economics: 56, 89-103.
[8] Jari L, Maijia S(2009). Hopping on one leg—The challenge of ecosystem disservices for urban green management. Urban forestry&urban greening. 8:309-315.
[9] Lai L, Huang X J,Wang H, Dong Y H, Xiao SS(2009). Estimation of environmental costs of chemical fertilizer utilization in China. Acta Pedologica Sinica. 46(1):63-69.
[10] Manfred K(2004). Green roof technoligy-from a fire-protechtion system to a central instrument in sustaionable urban design[A]. Greening rooftop for sustainable communities. Portland Conference[C]: 373-386.
[11] Ouyang Z Y, Wang X K, Miao H(1999). A primary study on Chinese terrestrial ecosystem services and their ecological-economic values.19(5): 607-613.
[12] Pang J(2008). Selection and valuation of city landscape plants tolerant to heavy metal contamination in Beijing. Beijing: Beijing Forestry University.
[13] Peng J, Wang Y L, Chen Y F, Li W F, Jiang Y Y(2005). Economic value of urban ecosystem services: a case study in Shenzhen. Acta Scicentiarum Naturalium Universitatis Pekinensis. 41(4): 594-604
[14] Primmer E, Furman E(2012). Operationalising ecosystem service approaches for governance: Do measuring, mapping and valuing entegrate sector-specific knowledge system. Ecosystem Service.1(1): 85-92.
[15] Sun M, Du X Y, Yang W R(2010). Evaluation model of plant landscape in park green space of Beijing and its application. Journal of Beijing Forestry University.32(Suppl 1): 163-167.
[16] Tang X H(2011). Evaluation of farmland shelterbelt ecological service function in northern Jiangsu—the case of Xinyi city. Nanjing, Nnajing Forestry University.

[17] Tyrvainen, L(2001). Econimic Valuation of Urban Forest Benefits in Finland. Journal of Environmental Management.62: 75-92
[18] Wang C J(2008). Research on estimating the ecological service value of the Chengdu central urban green space system. Ya'an, Sichuan Agricultural University.
[19] Wang W W, Dai Q C, Zhu M Y(2011). Ecological benefits of greening and related controlling factors in urban residential areas of Hangzhou: A quantitative analysis. Chinese Journal of Applied Ecology. 22(9): 2383-2390.
[20] Zhang X L, Xu Z J, Zhang Z H, Gu D Q, Wang L L(2011). Environment purification service value of urban green space ecosystem in Qingdao City. Acta Ecologica Sinica. 31(9):2576-2584
[21] Zhou S L(2007). Effect of N,P,K fertilizer on Tall Fescue growth and cold-resistance. Journal of Anhui Agri. Sci. 35(15):4467-4497.
[22] Zhou B B, Li Z K(2000). The value of forest resources of Beijing. Beijing: China Forestry Publishing.

A study on Spatial distribution of Chinese Traditional Villages based on GIS grid and remote sensing

Liang Yu , Xiaoli Meng
Gold mantis construction college
Soochow university
Dushu Lake Campus, Suzhou, Jiangsu
China
yuliang_163cn@163.com cateok@yeah.net

Abstract: China is a country with a long history of agricultural civilization, dotted with numerous ,various and historic traditional villages. Chinese traditional villages are rich in historical and cultural landscape, are the greatest legacy left by the Chinese agricultural civilization. Domestic scholars have done various studies on Chinese traditional villages. However , most scholars studied the traditional villages on the smaller scale. And there is no scholars apply the geography grid applications in research of traditional villages, the this study, we use the geographic grid methods, combined with GIS and remote sensing images to study the spatial pattern of the traditional villages ,including the list of three batches of the ancient village (total 2542), and the purpose is to understand the spatial distribution of Chinese traditional village better. We conclude that the spacial distribution of traditional villages in China is very uneven, and is extremely relevant to the population distribution, and the traditional villages are concentrated in I 9, I 13, I 14, I 15, I 18, I 19, I 20; and the example illustrate that the different levels of grid can make the distribution of the traditional villages more clear and reasonable.

Key-Words: Traditional villages Geographical grid Spatial distribution GIS Remote sensing

1 Introduction

Chinese traditional villages are rich in historical and cultural landscape, are the greatest legacy left by the Chinese agricultural civilization, China is a country with a long history of agricultural civilization, dotted with numerous ,various and historic traditional villages , in the process of formation of the traditional villages, both natural and human environment have significant impacts, traditional villages mix settlement pattern with natural environment ,settlement life and cultural[1]. The traditional villages surrounding environment can be divided into the natural environment, social and cultural environment, the impact of these factors on the performance of the traditional villages are in the traditional village of positional distribution, type, size and structure of the building. Village, mainly refers to the large group of settlements or some settlements formed a group, commonly used for expressing regional population concentration in the modern sense, including natural village, village area; settlement is a place for human living, it is not only a collection of construction, but also includes other living facilities and production facilities and, is directly related to urban settlements and rural settlements. In this study we use three batches of traditional villages announced by Chinese Housing and Urban Construction, Ministry of Culture and Finance, we study the relationship between the village and the environment based on GIS (geographic information system), especially analysis spatial distribution of traditional villages using GIS grid, figure out the positional relationship between villages , this study will be the basis of further studies of traditional villages .

Domestic scholars have done various studies on Chinese traditional villages, part of them concentrated on the cultural and tourism value of traditional villages and the development and protection of traditional villages[2]. Some scholars studied the ancient village space layout and geographical factors[3-4]. Other scholars introduced new ideas to the traditional village studies; there are also scholars studied the protection and development of ancient villages based on the spatial distribution[5-6]. In the study of spatial distribution of traditional villages, Yan analysis the characteristics of Chinese traditional villages distribution by the nearest neighbor index, including 1561 Chinese traditional villages[7]. Liu, who studied the spacial distribution of two batches traditional villages (total 1561), using the Moran's global index, local Moran's index and local association index[8]. Li et al studied the spatial distribution and influencing factors of ancient villages in Hunan province (total 101) based on GIS platform[9]. Domestic scholars have done various studies on Chinese traditional villages, part of them concentrated on the cultural and tourism value of traditional villages and the development and protection of traditional villages. In addition, there are also many researchers who studied geographic grid application. Dutton studied the global network model based on triangulation, and discussed the selection of the grid geometry model, data coding[10-11]. Some scholars evaluated the landscape, the landscape classification and ecological environment using humanities, economic, social, natural and other multi-source data based on a unified geographic grid[12]. These studies show that the geographic grid can be applied to different areas, providing a framework for the integration of multi-source data. However, most scholars studied the traditional villages on the smaller scale. And there is no scholars apply the geography grid applications in research of traditional villages, the this study, we use the geographic grid methods, combined with GIS and remote sensing images to study the spatial pattern of the traditional villages ,including the list of three batches of the ancient village (total 2542), and the purpose is to understand the spatial distribution of Chinese traditional village better.

2 Methods and Data

2.1 method of geographical grid classification

Mainland of China (except Hong Kong, Macao and Taiwan regions) is about 5200 kilometers east-west span, about 5500 km north-south span. In order to make the grid cover the entire mainland and to make grid-splitting reasonable, we split the grid into three different density levels, the density of villages in grid determine the level of grids. The grids having larger density in Southeast use three levels of grid, grids having smaller density in Northwest use the first stage grid .The size of the first stage grid is 1000km * 1000km, total 25; the second stage grid is based on the first, the size of the second stage grid is 500km * 500km, total 100; the size of the third stage grid is 250km * 250km, total 400 (Fig.1).

At the upper left corner of the grid as the starting point, the coordinates of the starting point of 66 degrees east longitude, 49 degrees north latitude, this point as the starting point of the grid can get entire continent (except Hong Kong, Macao and Taiwan) contains precisely . In order to make traditional villages statistical study conveniently ,we number from the top left corner of the grid sequentially, the primary grid are numbered Ⅰ1, Ⅰ2 ...... Ⅰ25, secondary grid numbered Ⅱ1, Ⅱ2 ... ... Ⅱ100, three grid numbered Ⅲ1, Ⅲ2 ...... Ⅲ400.

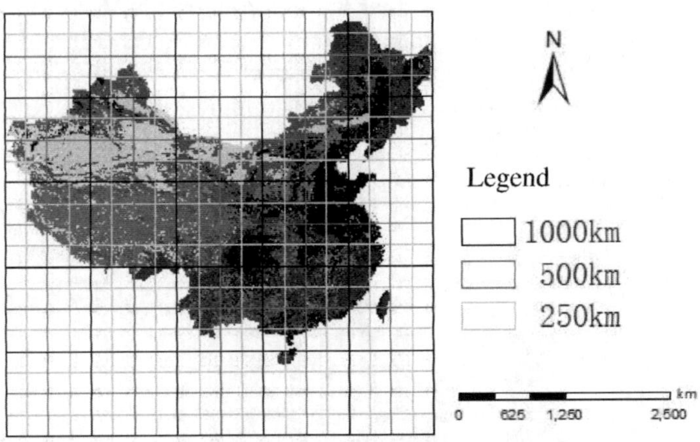

Fig.1 Geographical grid

2.2 method of geographical grid classification

In order to protect the traditional villages better, Housing and Urban Construction, Ministry of Culture and the Ministry of Finance of China (hereinafter referred to as the 3 sector) initiated and organized the selection of Chinese traditional villages, from 2012 until now has been released three batches (not included Hong Kong, Macao and Taiwan regions) , total 2542 traditional villages. In this study the data coverage continent-wide (Hong Kong, Macao and Taiwan range does not take into account), including three batches traditional villages data: the number of the first batch of national traditional villages directory published in December 2012 is 633, the number of the second batch of national traditional villages directory published in August 2013 is 915, the number of the third batch of national traditional villages directory published in November 2014 is 994, total 2542.

To give the traditional villages geographical coordinates, we obtain the geographical coordinates of the traditional villages (latitude and longitude) based on Baidu map. In accordance with "the Chinese Traditional Villages List", we record the geographical coordinates data. Combined with Baidu map coordinates-pickup function , we find that for part of the villages we can get latitude and longitude coordinates directly by Baidu map; for some villages we can get clearer satellite images , the methods to obtain geographical coordinates based on the obtained satellite images are as follows:

a. For a more balanced distribution of dwellings in a small village, we choose latitude and longitude coordinates of geometrical center;

b. For the larger villages, we choose latitude and longitude coordinates of traditional architecture geometrical center. (Fig.2)

Meanwhile, according to the above methods, there are some villages we can't obtain the latitude and longitude coordinates ,we choose latitude and longitude coordinates of town.

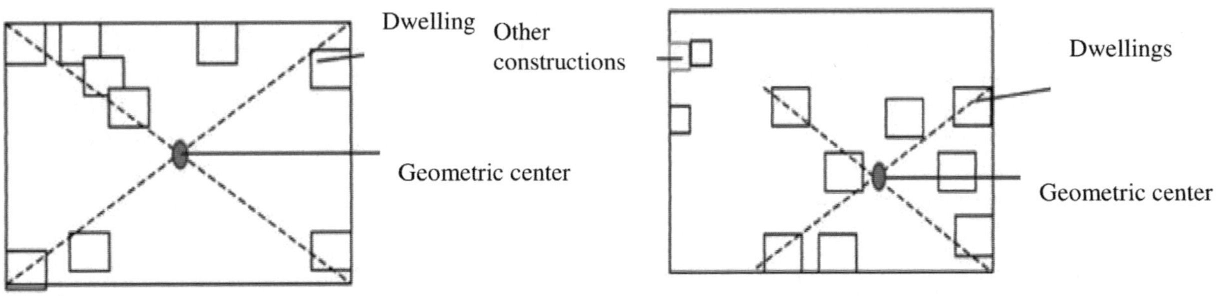

Fig.2 The method of obtaining the position coordinates of the points

We obtain the geographical coordinates of the traditional villages with the above methods and analysis the data statistically, the first batch of the villages can be more easily obtain the coordinates of the village according to the method described above accounted for 80%, accounting for 64% of the second batch, the third batch accounted for 82%, the rest of the village obtain the town coordinates (Fig3). In this research, we mainly study the nationwide spatial distribution of villages, with respect to small-scale internal research and villages, the spatial scale of this study is larger, so we can use a point to replace the village, for a small part of the data we use the town coordinates instead, and the affect in this study can be ignored.

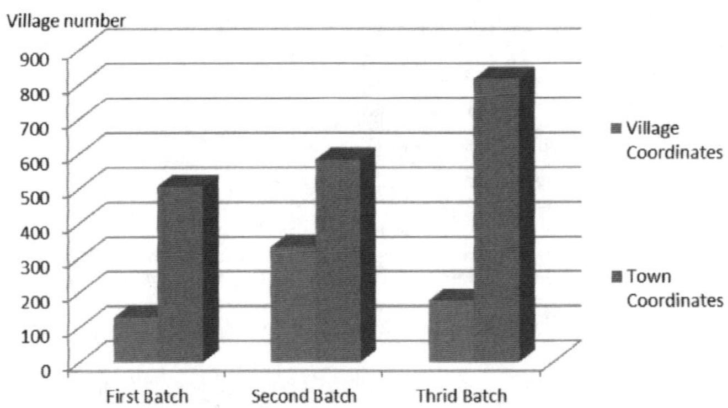

Fig.3 Traditional village coordinate statistical schematic

To visualize the spatial distribution of the traditional villages, we get Figure 4 based on ArcGIS software. As can be seen from the figure, spatial distribution of the traditional villages is very uneven, and the traditional villages are concentrated in southeast coast, southwest and central of China, and the number of traditional village in the Northwest and the Qinghai-Tibet region is very small. For a more rational analysis of the spatial distribution of traditional villages, we use the geographical grid method.

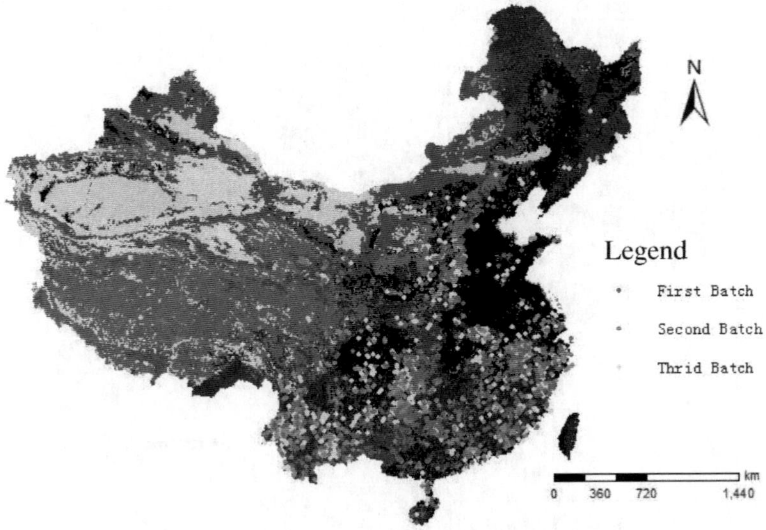

Fig.4 The spatial distribution of the traditional villages

According to National Geographic Grid (GB/T12409--2009), in this study the geographical grids are based on grid 2000 National Geodetic coordinate system, using Gauss - Krieger projection. The coordinates of this projection have rigorous mathematical relations with the latitude and longitude coordinates, can be transformed by Gaussian positive and negative calculation formula, and in our country the topographic maps are also based on this coordinate system, so if we use the coordinate system, there will be a large number of basic survey data, broad application.

3 Spatial distribution analysis of traditional village based on geographical grid

3.1 Spatial distribution analysis of traditional village based on geographical grid

In order to analyze the quantity and distribution of traditional village in the geographical grid, we analyze the traditional village in the first level grid , the villages are mainly concentrated in Ⅰ9, Ⅰ13, Ⅰ14, Ⅰ15, Ⅰ18, Ⅰ19, Ⅰ20 grid. In the Ⅰ19 ,the villages accounted for 31% of the total number of traditional villages (Fig.5), we will take this region as an example to analyze the village distribution in the next level grid (Fig.6).

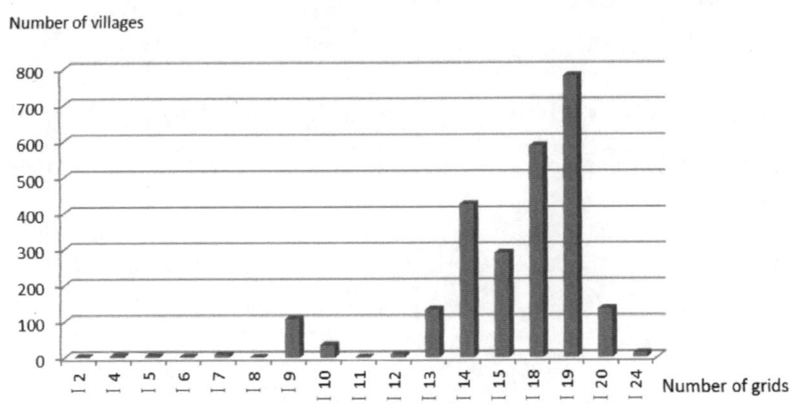

Fig.5 Statistics data in the first level grid

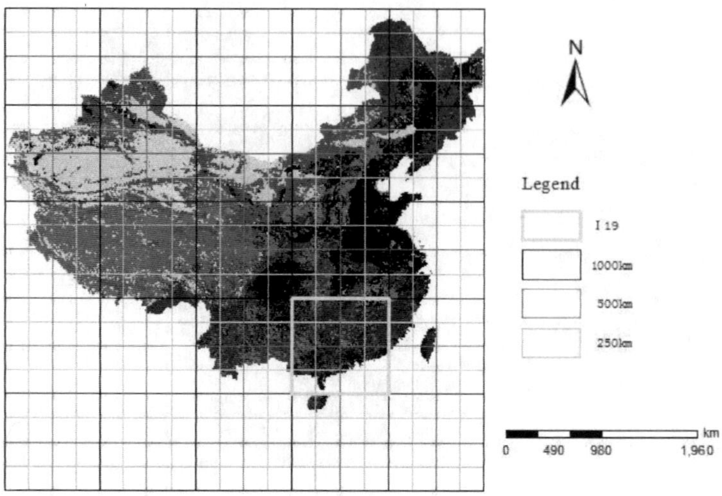

Fig.6 The location of Ⅰ19

In order to analyze distribution of the traditional village of the second stage grid ,we extract the data in Ⅰ19, we find that the Ⅱ67, Ⅱ68, Ⅱ77, Ⅱ78 within this range, and the village distribution as shown in Fig.7. The villages located in Ⅱ67 accounted for 66% of the total number of villages in Ⅰ19 , so it is necessary to analyze the village distribution of next level .

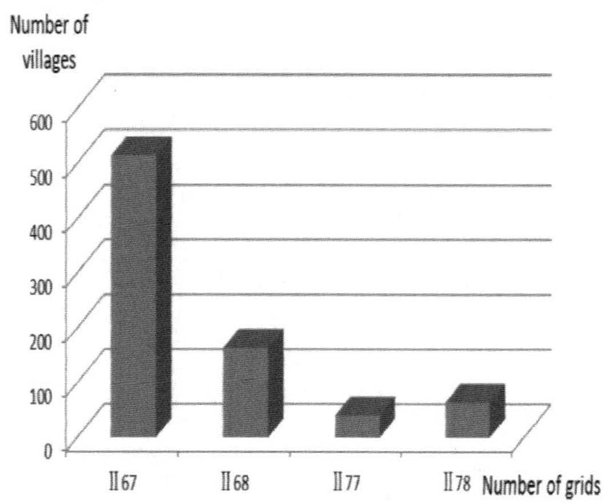

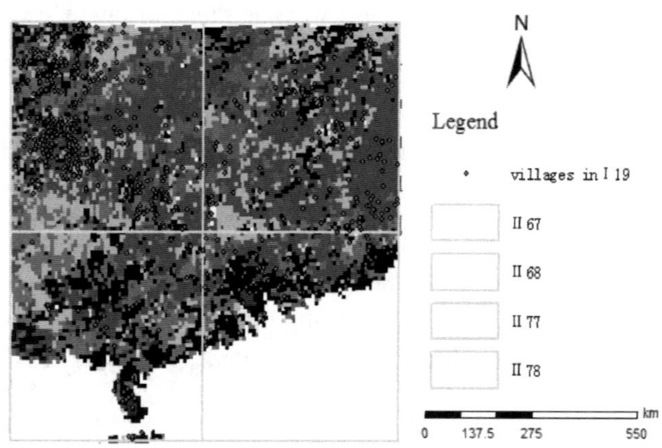

Fig.7.a The distribution of villages in Ⅰ19 **Fig.7.b** The distribution profile of villages in Ⅰ19

The Ⅲ253, Ⅲ254, Ⅲ273, Ⅲ274 are located in Ⅱ67,and there are 167,27,241,79 villages in these grids, the spatial distribution is shown in Fig.8.

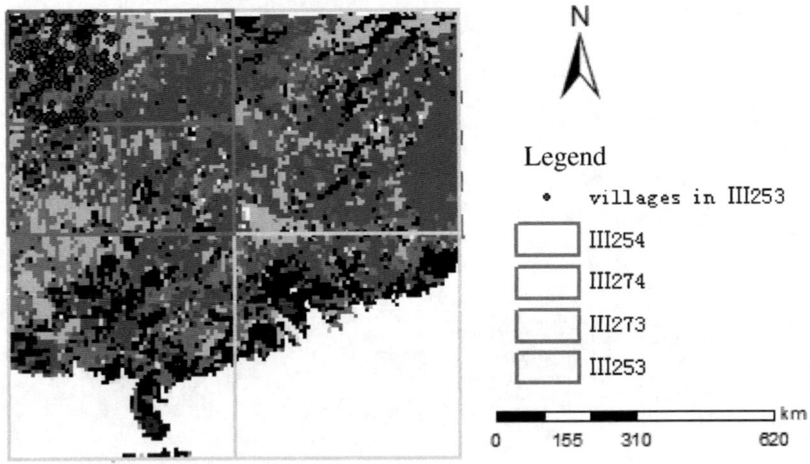

Fig.8 The distribution profile of villages in II 67

3.2 Discussion

By the grade geographic grid, we can easily calculate the number of traditional villages in a certain grid, and have a clear understanding of the spatial distribution pattern of the villages. At the same time, we can integrate aspects of data about villages effectively through the grid. In addition, the method setting small grid in the large grid can store data more reasonablely and reduce the waste of resources. Taking the I 15 as an example, the III 237 in the third stage grid has 191 villages ,combining with remote sensing images referred to above. Huanxi Village , Jiangnan Town in Hangzhou(Fig.9-a), Yaoxi Village, Hecun Town in Hangzhou (Fig.9-b), Maoping Village ,Fuchunjiang Town in Hangzhou (Fig.9-c), Qinzhou Village, Langchuan Town in Hangzhou (Fig.9-d) both are located in III237 .Based on the geographic grid combined with remote sensing data, we can quickly perform a variety of data integration, obtain more comprehensive information of villages and the surrounding environment of villages, such as density of the villages , the characteristics of single building in villages, the relationship between traditional and modern urban development, and the villages surrounding natural landscape and roads. Of course, due to the economic factors, we can't obtain all of remote sensing image of every village through public network resources easily. In general, villages that are vast, relatively remote from the city, are limited to obtain remote sensing images or only get low resolution images, and are unmanageable to analyze the relationship between the village and the surrounding environment and roads. So the value of data is low, we need another approach to obtain data.

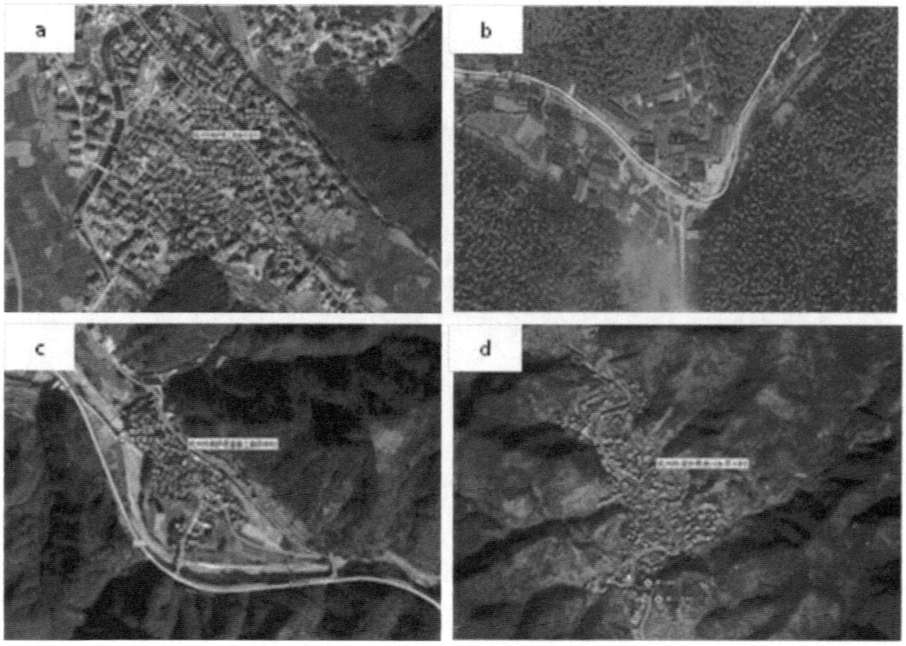

Fig.9 Remote sensing image of villages

4 Conclusion

In this study, we analyze the spatial distribution of three batches traditional villages that has been published by Housing and Urban Construction, Ministry of Culture and the Ministry of Finance of China based on GIS geographic grid method ,and we conclude that the spacial distribution of traditional villages in China is very uneven, and is extremely relevant to the population distribution, and the traditional villages are concentrated in Ⅰ9, Ⅰ13, Ⅰ14, Ⅰ15, Ⅰ18, Ⅰ19, Ⅰ20; and the example illustrate that the different levels of grid can make the distribution of the traditional villages more clear and reasonable. So we conclude that , we can study the relationship between the villages better based on the geographic grid ,and the geographic grid provides spatial distribution characteristic of villages. Based on the method that low village density area with large-scale grid, high village density area with small-scale grid, we can effectively study the relationship between the formation of the traditional villages and geography. However, the data of Chinese traditional village is not complete, we can obtain few data through public way .For our study , we only use three batches of traditional village as well as some remote sensing data. In the future, we hope that we can obtain more data to analyze the spatial distribution pattern of the traditional villages and broad the research field. In addition, in our study, the grid are numbered, but some of the grid don't encompass the villages, and therefore there are issues involved in the active grid, in the further study, we will carry out a detailed analysis of this issue, in order to improve the operation efficiency of data, to save storage space.

This research Supported by National Natural Science Foundation of China:The Primary Research on Geospatial Distribution Law of Chinese Traditional Dwellings (No.41371173).

References:

[1] Xiaoxing Liu, Organic Evolution and Enlightenment of Chinese Traditional Settlement, *Urban Planning Forum*, No.11, 2007, pp. 55-60.

[2] Weiming Tan, Review of Ancient Village Tourism in China, *Journal of Guizhou institute of education (social science)*, Vol.25, No.11, 2009, pp. 48-51.

[3] Jing Zhang, The Relationship Between space distribution and Topography degree in Xinzhou rural settlement, *Journal of Xinyang normal university (natural science edition)*, Vol.28, No.2, 2015, pp. 195-198.

[4] Daodian Lu, The formation and development of ancient villages and space distribution characteristics in Zengcheng,*China Ancient City*, 2014, pp. 63-67.

[5] Xiaofei Hou, Spatial distribution of tourism development and protection of ancient villages in Shanxi Province, *Journal of Shanxi normal university (natural science edition)*, Vol.28, No.4, 2014, pp. 112-115.
[6] Lei Tang, The characteristics of spatial pattern protection and industrial development research of three ancient villages in Suzhou Xishan, *Urban Planning International*, Vol.24, No.2, 2009, pp. 112-116.
[7] Sai Yan, Characteristics and Cause Analysis of Chinese Traditional Villages Distribution, *Journal of Dali University*, Vol.13, No.9, 2014, pp. 25-29.
[8] Dajun Hu, The study of Chinese traditional villages spatial distribution pattern, *China Population Resources and Environment*, Vol.24, No.4, 2014, pp. 157-162.
[9] Bohua Li, The spatial distribution characteristics and influencing factors analysis of traditional villages in Hunan province, *Economic Geography*, Vol.35, No.2, 2015, pp. 189-194.
[10] Dutton, Polyhedral hierarchal tessellation: the shape of GIS to come, *GeoInfo Systems*, Vol.3, No.1, 1991, pp. 49-51.
[11] Dutton, Improving locational specialty of map-Amulyi-resolution, metadata-driven approach and natation, *IJGIS*, Vol.10, No.2, 1996, pp. 251-268.
[12] Chenghu Zhou, Progress in Geography Grid Model, *Progress In Geography*, Vol.28, No.5, 2009, pp. 657-662.

Climate change adaptation and mitigation measures in riverine systems

PETER HALAJ[1], VILIAM BAREK[1], DUSAN IGAZ[2], JAN HORAK[2]
[1]Department of Landscape Engineering, [2]Department of Biometeorology and Hydrology
Slovak University of Agriculture in Nitra
Hospodarska 7, 94976 Nitra
SLOVAKIA
peter.halaj@uniag.sk http://www.spu.sk

Abstract: Human impact on aquatic ecosystems causes that surface waters often suffer from significant hydromorphological and hydrological alterations and changes in adjacent floodplains. These changes impact the water body status and therefore need to be addressed in the near future. Land use in stream basin often exhibited high loads of sediments and surface runoff that affect stream water quality. It leads to collapse of stream biota and permanent eutrophication caused by excess nutrient inputs. Generally, main efforts are under way to restore watersheds, but restoration planning rarely accounts for projected climate change. There is indicated a large negative impact of climate change on freshwater habitats. Therefore, restoration and protection help to mitigate the adverse effects and may enable populations to survive in conditions of climate change. The authors analyse impacts of climate change on streamflow hydrology, present projected changes of climate conditions in Slovakia and their impact on streams and summarize principles of channel-floodplain geosystem restoration that enables to approach to the state of dynamic stability as a basic assumption for increasing of „absorbing capacity" of freshwater ecosystems in climate change conditions.

Key-Words: climate change, river restoration, stream ecosystem, flow regime, riverine system, adaptation and mitigation measures

1 Introduction

The continuing multiple anthropogenic stressors to streams, reduce the natural resilience and resistance capacity of freshwater ecosystems that enables them to recover from certain levels of perturbation [10]. A physical habitat and ecological processes in natural systems are often significantly modified by human activities [10]. Although, some of the anthropogenic alterations provide important benefits for the human population, they often degrade elements of freshwater ecosystems, decline in biodiversity results largely from habitat modifications. There is also an urgent need for understanding of the implications of climate change for freshwater ecosystems. Compared to terrestrial ecosystems, processes in aquatic ecosystems and their conservation, particularly in stream habitats, are still poorly understood and offer a great perspective for future research and conservation [8]. Successful sustainable rehabilitation of the streams is depending on the identification and understanding processes and anthropogenic induced changes that enable to determinate an appropriate longitudinal and lateral riverine system structure including stream habitat improvements at different scales. Key consideration must be given to whether changes to the boundary conditions under which the river operates have brought irreversible changes to river structure and function [6].

2 Impact of Climate Change on streamflow hydrology

The natural variability fluctuations of freshwater ecosystems and water supplies are governed by the climatic regime of a region and the prevailing hydrologic cycle. The vulnerability of freshwater ecosystems to hydrologic variability may constraints ecological and evolutionary processes in streams [16], [21]. Bunn and Arthington [3] and Macura et al. [18] state that flow is a key factor of hydromorphological quality of habitat in streams, which in turn is a major determinant of biotic composition. Droughts put aquatic biota under increasing stress as they represent progressive loss of aquatic habitat, depletion of food resources and decline of water quality, with an increased likelihood of biotic interactions as flow reduces [2].

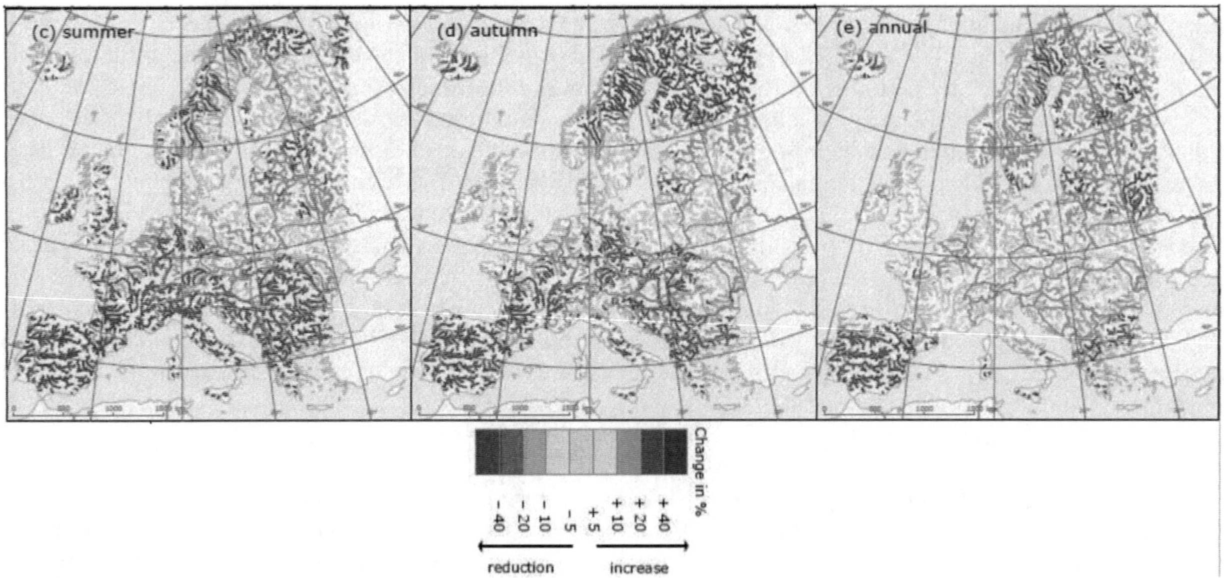

Fig.1 Projected change in mean seasonal and annual river flow between 2071–2100 and the reference period 1961–1990 (Adopted from [4]).

The above-mentioned facts must be taken into account in the context of climate change impact on streamflow patterns (Fig. 1). Strong changes are predicted for the seasonal pattern of river flows, with distinct differences in Europe. Annual river flow is projected to decrease in southern and south-eastern part of Europe and increase in northern and north-eastern Europe [23]. Winter and spring river flows are projected to increase in most south-eastern regions. In summer and autumn, river flows are projected to decrease in most of Europe, except for northern and north-eastern regions [4].

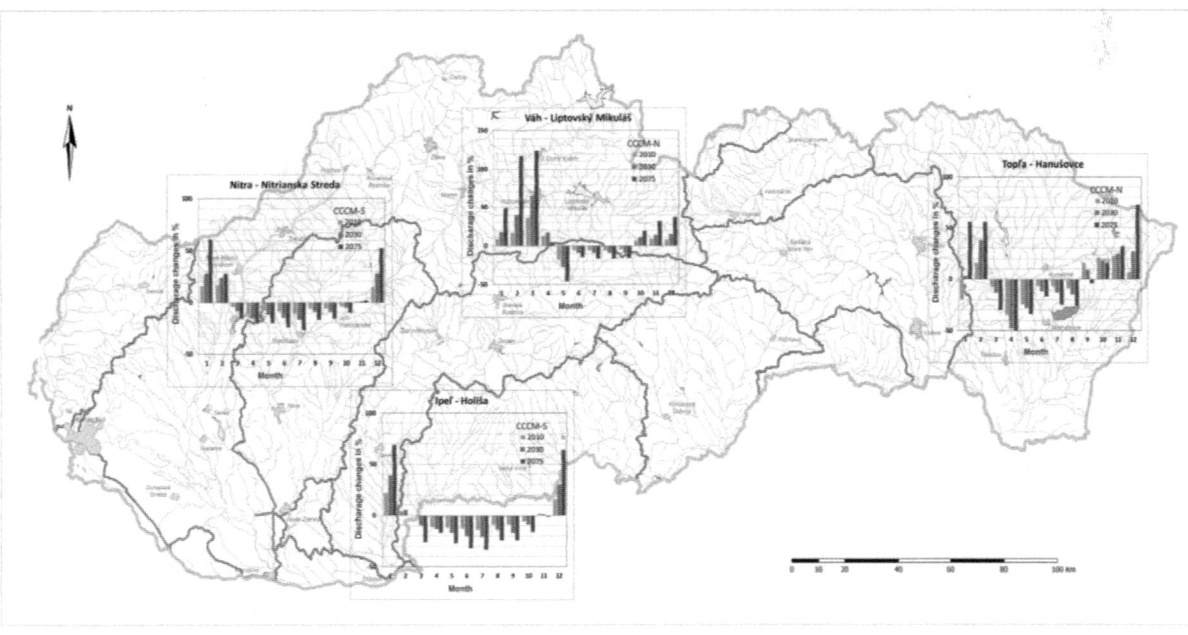

Fig.2 Changes of mean monthly discharges for time horizons 2010, 2030 and 2075 according to regional outputs of CCCM.

According to an evaluation of hydrological scenarios of seasonal stream flow distribution changes for time horizons 2010, 2030 and 2075 for CCCM model in Slovakia (Fig. 2), we may state that from the north to south the runoff increase in winter period is shortened. The maximum runoff increase is usually occurring in January or February. The highest discharge decline is projected for the spring months in the northern basins, the decline shifts for the summer months in southern basins. We should expect discharge losses for the greater part of a year that increase when considering southern regions of Slovakia. The highest discharge decline occurs in the spring months and during the autumn months runoff decline increasing is frequent [13], [24]. A more frequent occurrence of extreme events can be expected, in particular, flash and urban floods, triggered by local intensive precipitation events [14]. The spatial analysis of minimal discharges for the period 1961 – 2012 for Slovak regions divided into catchments with various vulnerabilities of minimal discharges show downward trend in southern and south-western part of Slovakia except the Danube interbasin. The upward trend is projected for eastern and north-eastern parts of Slovakia. The downward trends prevail in May, June, July and in some areas in the autumn-winter period [20].

3 Projected Changes of Climate conditions in Slovakia and their impact on streams

The situation is made more difficult by projected changes in climate conditions in Slovakia. Figure 4 and 5 Projected changes of monthly mean precipitation (for Southern Slovakia) and monthly mean air temperature (average values for Slovakia) according to regional outputs of CCCM. The climate change scenarios for time horizons 2010, 2030 and 2075 according to the CCCM for Slovakia. Lapin et al. [15] show a moderate decrease and increase in precipitation and increase of air temperature for all the time horizons. A gradual decreasing of precipitation from May to September for both northern and southern parts of Slovakia are projected (figure 4 and 5).

Gaál et al. [7] examine projected changes in precipitation extremes in simulations of 12 regional climate models (RCMs) for Central and Southeastern Europe projected for time slice 2070 – 2099). They state that drying is projected for seasonal totals in all RCMs while increases clearly prevail for the 50-year return values. The projected increases are larger for short-term extremes, mostly related to convective events in comparison with multiday extreme rainfall events.

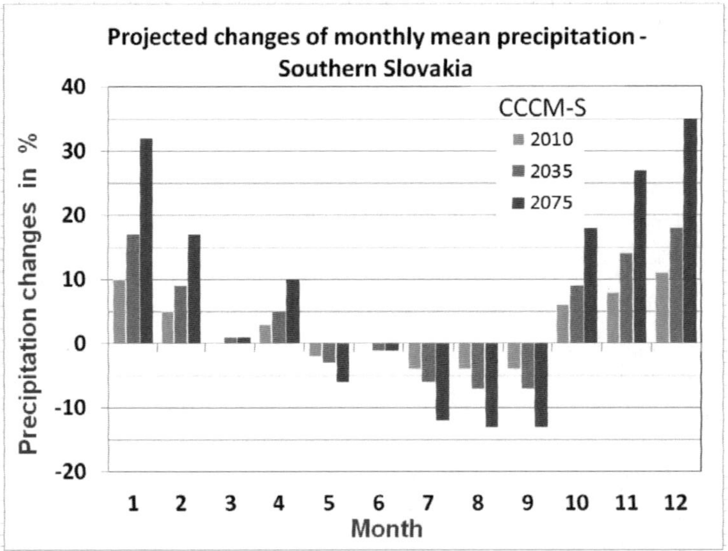

Fig.4 Projected changes of monthly mean precipitation (for Southern Slovakia) according to regional outputs of CCCM.

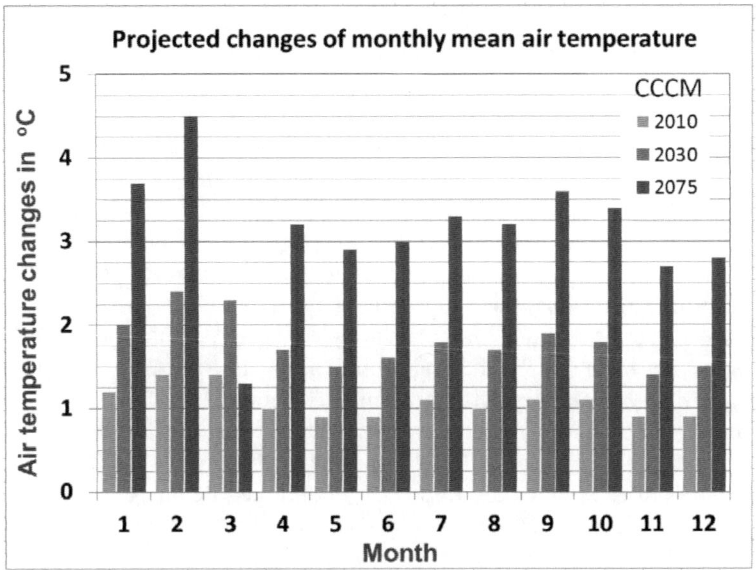

Fig.5 Projected changes of monthly mean air temperature (average values for Slovakia) according to regional outputs of CCCM.

These changes will have negative consequences on riverine systems by influencing abiotic and biotic conditions which diminish natural variability and threaten the original ecosystem and they are summarized in Table 1.

Table 1 Impact of Climate Change on streams

Climate-Driven Change	**Effects on Stream**	**Expected Response – abiotic components**	**Expected Response – biotic components**
Decreased spring and summer discharges	Decreased water depth	Lost of refugia Stream habitats fragmentation;	Decreased abundance and diversity of aquatic fauna, Change of riparian vegetation, Decreased biological activity in hyporheic zone
	Decreased mean velocity	Changes of physical habitat Increased concentration of contaminants	Changes in composition of aquatic fauna Excessive growths of submerged aquatic macrophytes
	Streambed drying	Interruption of exchange between surface and subsurface water	Changes in composition of aquatic fauna in hyporeic zone
Increased summer air temperatures	Increased water temperature	Decreased dissolved O2 Decreased water quality Decreased self- purification ability	Changes in composition of aquatic fauna Occurrence of invasive species
Decreased cloudiness	Increased solar radiation exposure	Water overheating	Excessive growths of submerged aquatic macrophytes
More frequent occurrence of storm events	More frequent occurrence of floods	Physical habitat destruction	Changes in composition of aquatic fauna
	Intensive erosion washes	Siltation of bed	Decreased spawning success Decreased activity of hyporheic invertebrates

4 Climate Adaptation and Mitigation measures in riverine system

An elimination of climate change resulted in decrease of annual mean discharge (- 42.4 % for time horizon 2075 according to CCCM in comparison with reference period (1931-1980) for territories with elevation below 300 m a.s.l. [24], where the most intensively cultivated areas of Slovakia are located, require appropriate adaptation measures. The necessity to take effective measures is also given by the projected increase of irrigation water need. According to the CCCM projected irrigation water need for time horizon 2075 equals about 178 % of amount for time horizon 2010 [1]. There will probably exist a capacity limit of environmentally friendly mitigation measures to cover strong increase of water demand. In such cases, one of the possible alternatives is building of small reservoirs in water-limited catchments. In the past, an expansion of modern dam-building started in the mid-20th century but recognition of negative impacts of these structures on riverine ecosystems over the last decades have caused that dam removal has become an increasingly common activity in river restoration projects [9]. River reaches with streamflow regulated by dams loss active floodplain zones [11]. It causes a decrease of biodiversity patterns variation related to the complex hydro-geomorphology of such systems and the connectivity changes through space and time [17], [27]. In spite of the frequently mentioned negative Influences of In spite of the frequently mentioned negative influences of dam construction, they represent one of the effective measures for mitigation of climate change impacts covering increases of water demand. Poff and Matthews [19] state that 'advances in scientific understanding and technical tools over this period contributed to shifting conservation and social expectations for how dams might be better designed and operated for diminishing adverse ecological impacts'. For mitigating negative ecological impacts of dams, relevant measures should be adopted based on optimisation of spatial arrangement of dams and human induced changes in hydrographic network [17], changing of operation rules to convert steady flow regime to the more natural one [21], restoration of natural depressions with capacity for storing water during high flows, restoration of tributaries located below dam, restoration of reaches located below dam, restoration of wetlands and reassessment of environmental flows to mitigate the undesirable hydrological impacts of dams [19] and design of fish-passage facilities to maintain healthy runs of migrating species and improvement of downstream water quality by aeration and temperature modification [12]. River restoration strategies are generally focused on restoration of ecosystem functions to a level that meets ecological and socioeconomic objectives [5]. The restoration projects are compared to pre-restoration conditions and historic, natural conditions, both assumed as boundaries of the worst and the best state of a river system [22]. Successful projects require using a holistic approach means regarding river systems as interrelated environmental, social and economic systems integrated into the green infrastructure of rural and urban landscapes and they should be designed as a multi-purpose project addressed also to climate change impact mitigation and ecosystem service provision. A well-known set of the basic principles of river restoration has been formulated by the Watershed Ecology Team [25]. These principles define a basic framework focused on scientific and technical issues of the river restoration process. Focusing on the restoration of natural processes avoids the misapplication of restoration techniques by enabling the natural array of habitat types to form in all parts of a stream network. Moreover, this approach provides suitable habitats for all native aquatic species because it restores the conditions to which local fish stocks are adapted. Thus, it avoids the problem of building habitats that may improve habitat for one species yet degrade the habitat. Ecosystems are moving targets, with multiple potential futures that are uncertain and unpredictable. Therefore, the management has to be flexible, adaptive, and experimental at scales compatible with the scales of critical ecosystem functions [26].

4 Conclusion

The implementation of Water Framework directive requires an analysis of current state of freshwater ecosystem and human activities impact assessment that is necessary for integral water management application and appropriate measures design for aquatic and terrestrial zones of stream ecosystem. To fulfil river restoration goals we have to change the traditional engineering approach to river management and developing a practical approach to adaptive and ecosystem based management with respect to carrying capacity of water resources. To contribute more effectively to sustainable freshwater management, we must focus not only on restoration but also on adaptation to climate change and other environmental challenges.

Acknowledgements:
This work has been supported by the Research Support Scheme of the VEGA project no. 1/0744/13.

References:

[1] Bárek, V., Halaj, P., Igaz, D., The Influence of Climate Change on Water Demands for Irrigation of Special Plants and Vegetables in Slovakia, In *Bioclimatology and Natural Hazards*, Dordrecht : Springer, 2009, pp. 271-282.

[2] Bond , N.R., Lake, P. S., Arthington, A.H., The impacts of drought on freshwater ecosystems: an Australian perspective', *Hydrobiologia*, Vol. 600, No.1, 2008, pp.3–16.

[3] Bunn, S. E., and Arthington, A. H., Basic principles and ecological consequences of altered flow regimes for aquatic biodiversity. *Environmental Management*, Vol. 30, 2002, pp. 492-507.

[4] EEA, *Impacts of Europe's changing climate - 2008 indicator-based assessment*, EEA Report No 4/2008, 2008. ISSN 1725-9177.

[5] FISRWG, *Stream Corridor Restoration: Principles, Processes, and Practices*. By the Federal Interagency Stream Restoration Working Group (FISRWG)(15 Federal agencies of the US gov't). GPO Item No. 0120-A; SuDocs No. A 57.6/2:EN 3/PT.653. 1998. ISBN-0-934213-59-3.

[6] Fryirs, K., Guiding principles for assessing geomorphic river condition: Application of a framework in the Bega catchment, South Coast, New South Wales, Australia, *Catena* Vol.53, No.1, 2003, pp.17–52.

[7] Gaál, L., Beranová, R., Hlavčová, K., Kyselý, J., Climate Change Scenarios of Precipitation Extremes in the Carpathian Region Based on an Ensemble of Regional Climate Models', *Advances in Meteorology*, Vol. 2014, Article ID 943487, 2014, p.14, http://dx.doi.org/10.1155/2014/943487

[8] Geist, J., Integrative freshwater ecology and biodiversity conservation, *Ecological Indicators* ,Vol.11, 2011, pp. 1507–1516.

[9] Gerlak, A.K., Susanna Eden, S., Megdal,S., Lacroix, K.M. Schwarz, A., Restoration and river management in the arid southwestern USA: exploring project design trends and features', *Water Policy* 11, 2009, pp.461–480.

[10] Giller, P. S., River restoration: seeking standards. Editors introduction, *Journal of Applied Ecology*, Vol. 42, 2005, pp. 201-207.

[11] Graf , W.L., Downstream hydrologic and geomorphic effectsof large dams on American rivers. *Geomorphology*, 79, 2006, pp. 336–360.

[12] Higgins JM, Brock WG., Overview of reservoir release improvements at 20 TVA dams, *Journal of Energy Engineering*, 125, 1999, pp. 1-17.

[13] Hlavčová K. and, Čunderlík, J. (1998) 'Impact of climate change on the seasonal distribution of runoff in mountainous basins in Slovakia', In: *Hydrology, Water Resources and ecology in Headwaters (proceedings of the HeadWater'98*, Conference held at Merano, Italy, April 1998. IAHS Publ. No. 248, pp. 39-46.

[14] Kundzewicz , Z.W, Luis José Mata. L.J., Arnell, N., Döll, P., Kabat, P., Jiménez, B., Miller, K., Oki, T., Zekai Sen, Z., Shiklomanov, I., Freshwater resources and their management', *In Climate Change 2007: Impacts, Adaptation and Vulnerability. Contribution of Working Group II to the Fourth Assessment Report of the Intergovernmental Panel on Glimate Ghange*, Eds. M.L. Parry et al. (Cambridge, England: Cambridge University Press), 2007, pp. 173–210.

[15] Lapin, M., Nieplová, E., Faško, P., Regional scenario of air temperature and precipitation changes in Slovakia (in slovak): *Proceedings National Climate Programme of the Slovak Republic No.3*, ME SR, SHMI Bratislava, 1995, pp. 17–57.

[16] LeRoy Poff, N.L and Ward, J.V., Implications of Streamflow Variability and Predictability for Lotic Community Structure: A Regional Analysis of Streamflow Patterns. *Canadian Journal of Fisheries and Aquatic Sciences*, , 46(10),1989, pp. 1805-1818, 10.1139/f89-228

[17] McCluney, K.E., Poff, N LeRoy, Palmer, M.A., Thorp, J.H., Poole, G.C., Williams, B.S., Williams, M. R., Baron, J.S., Riverine macrosystems ecology: sensitivity, resistance, and resilience of whole river basins with human alterations ', *Front Ecol Environ,* 12(1), 2014, pp. 48–58.

[18] Macura, V. , Škrinár, A. Kalúz, K. Jalčovíková, M., Škrovinová, M., Influence of the morphological and hydraulic characteristics of mountain streams on fish habitat suitability curves, *River Research and Applications*, Vol. 28, No. 8, 2012, pp. 1161-1178.

[19] Poff, N.L.R, Matthews, J.H. Environmental flows in the Anthropocence: past progress and future prospects, *Current Opinion in Environmental Sustainability*, 5, 2013, pp.667–675.

[20] Poórová, J., Škoda, P., Danáčová, Z., Šimor, V. Evolution of Hydrological Regime of Slovak Rivers', (in slovak), *The Environment*, Vol. 47, No.3, 2013, pp.144 – 147.

[21] Richter, B.D. and Thomas, G.A., Restoring environmental flows by modifying dam operations, *Ecol Soc 12*, 2007, pp.12–38.
[22] Rohde, S., *River Restoration: Potential and limitations to re-establish riparian landscapes. Assessment & Planning*. Swiss Federal Institute of Technology, Zurich, Switzerland, 2004, 127 pp,
[23] Schneider, C., Laizé, C. L. R., Acreman, M. C. , Florke, M., How will climate change modify river flow regimes in Europe? ‘ *Hydrol. Earth Syst. Sci*., 17, 2013, pp. 325–339.
[24] Szolgay, J., Hlavčová, K., Parajka, J., Čunderlík, J., Assessment of the potential impact of climate change on river runoff in Slovakia’, Publication of the Slovak National Climate Programme of the Slovak Republic, No. 6. SHMI, ME SR, Bratislava, 1997.
[25] USEPA, Principles for the Ecological Restoration of Aquatic Resources’, EPA841-F-00-003. Office of Water (4501F), United States Environmental Protection Agency, Washington, DC, 2000, p. 4.
[26] Walters, C. J., *Adaptive Management of Renewable Resources.* Macmillan, New York, 1986.
[27] Ward, J. V. and Stanford, J.A., The serial discontinuity concept: Extending the model to floodplain rivers ’ , *Regulated Rivers: Research and Management*, Vol.10, 1995, pp. 159–168.

Temporal/spatial distribution of fine particulates and their formation mechanism

KANG ZHOU, JIAN HOU, ZHIZENG XIA, QINGLIANG WANG
School of Petroleum Engineering
China University of Petroleum
No. 66, Changjiang West Road, Huangdao District, Qingdao, 266580
CHINA
zhoukang_upc@163.com

Abstract: Fine particulates seriously affect environmental quantity, agriculture production, animal husbandry and human health. Based on Air Quality Index (AQI) data of 13 monitoring sites in Xi'an City since 2013, the paper studies the temporal/spatial distribution of fine participates and then analyzes how these distribution characteristics form. Results show that the AQI values of fine particulates are highest in winter (December, January, February) and lowest in summer (June, July, August). As for spatial distribution, the fine particulates mainly gather at the northern part of Xi'an City where industrial districts and populated area located. These temporal/spatial distribution characteristics result from the comprehensive effect of precipitation, temperature, weather, wind direction, terrain, industrial layout and population distribution.

Key-Words: fine particulates; correlation analysis; temporal/spatial distribution; formation mechanism

1 Introduction

Fine particulate matter refers to the particles with aerodynamic equivalent diameter less than or equal to 2.5 microns. They are commonly mentioned as PM2.5. The studies all over the world have shown that PM2.5 can reduce atmospheric visibility, cause traffic accidents, affect agricultural production, aggravate climate anomalies, and seriously harm human health[1-7].

Considering the negative effect of PM2.5, the United States Environmental Protection Agency (EPA) set the pollution standard for PM2.5 in 1997[8]. However, in China, only sulfur dioxide (SO2), nitrogen dioxide (NO2) and inhalable particles (PM10) were daily monitored before 2013. PM2.5 together with carbon monoxide (CO) and ozone (O3) was just added into the daily monitoring pollutants in 2013[9]. As a result, there is not useful historical data of PM2.5 for a long time in China, which limits the study on temporal/spatial distribution of PM2.5 and their formation mechanism.

Based on the Air Quality Index (AQI)[10] values obtained from 13 monitoring sites in Xi'an City since 2013[11], the paper studies the temporal as well as the spatial distribution of PM2.5 and then analyzes the main mechanisms behind these distribution characteristics. These works are very important for further research on PM2.5 and future design of industrial layout in more cities all over the world.

2 Temporal/spatial distribution of PM2.5 and their formation mechanism

In order to prevent the generation and reduce the harm of PM2.5, it is important to know more about their temporal/spatial distribution and formation mechanism. This paper carries out a case study of Xi'an City.

2.1 Average evaluation index

The monitoring data of PM2.5 has been provided every day in Xi'an City since 2013. So the amount of monitoring data is huge and the volatility is strong. In order to study the temporal/spatial distribution effectively, an average evaluation index is necessary. The statistical median is used in this paper. It reflects not only the particular AQI value but also the frequency of occurrence.

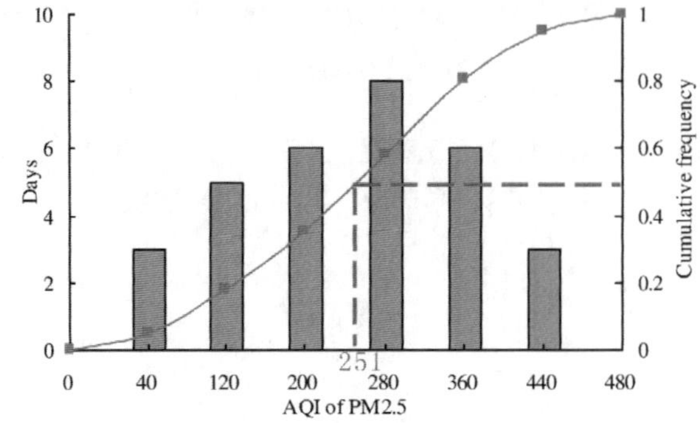

Fig. 1 Calculation of AQI statistical median in January

Fig. 1 shows the calculation process of statistical median of January in Xi'an City. First, distribute the AQI values into several subsets based on their distribution scope. Then, count the number of AQI values in each subset and draw the cumulative frequency diagram. Finally, calculate the statistical median under the cumulative frequency of 0.5 from the diagram.

2.2 Temporal distribution and its formation mechanism

Fig. 2 compares the AQI statistical medians of 12 months. Obviously, winter (December, January and February) has the largest AQI statistical medians of PM2.5 and they are all higher than 200. Summer (June, July and August) has the smallest AQI statistical medians of PM2.5 and they are no larger than 70. In order to study the main mechanisms behind these temporal distribution characteristics, the daily meteorological data such as weather, wind direction, wind power and temperature of each month are analyzed, as shown in Fig. 3(a-d).

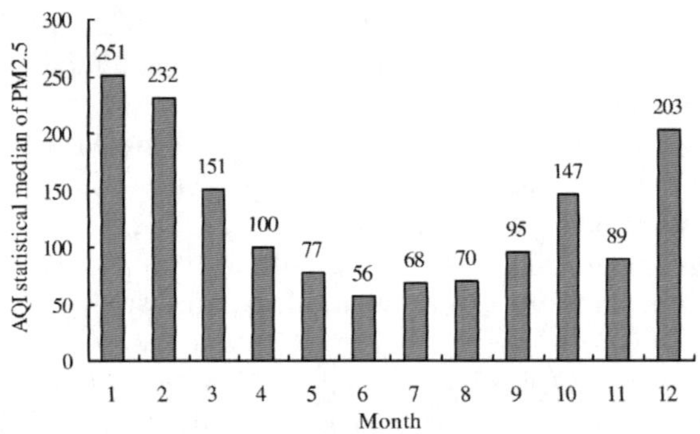

Fig. 2 AQI statistical medians of 12 months

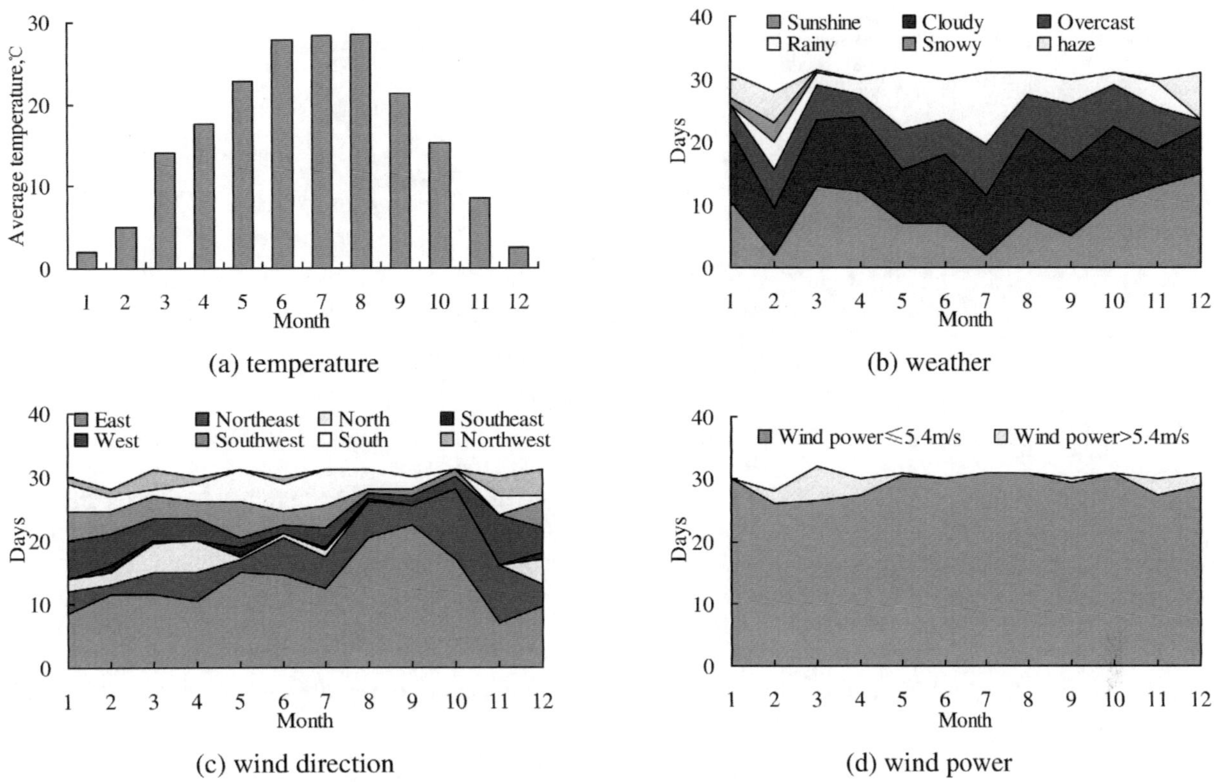

Fig. 3 Meteorological statistical data of 12 months

As can be seen from Fig. 3(a), the temperature is very low in winter. As a result, much more coal is burned and thus more PM2.5 is generated. Most plants wither in winter and thus their self-purification ability is lost. This also causes that dry soil is exposed in the air and generates some PM2.5. It is also easier to form thermal inverse layer above the city in winter, which affects the diffusion of PM2.5. Fig. 3(b) shows that the precipitation is relatively less in winter and some urban water also freezes. So the water self-purification ability is negligible. From Fig. 3(c-d), we can see that the west/southwest and north wind is relatively more in winter than other months. These winds can blow PM2.5 generated from industrial districts and loess plateau in northern and western Xi'an City to the whole city. All of these mechanisms contribute to the highest AQI statistical medians of PM2.5 in winter (December, January, February).

However, as can be seen in Fig. 3, the temperature is higher and the precipitation is more in summer. So the coal combustion is decreased and the plants cover the soil, which reduces the generation of PM2.5. The self-purification ability of water, air and plants is also increased. The high surface temperature also promotes cross-ventilation and it is useful for the diffusion of PM2.5. In addition, east/northeast wind is much more in summer, which blows PM2.5 away from the city. These mechanisms result in the lowest AQI statistical medians of PM2.5 in summer (June, July, August).

2.3 Spatial distribution and its formation mechanism

Fig. 4 summaries the locations of 13 monitoring sites in Xi'an City and their AQI statistical medians of PM2.5 all over the year. As can be seen, the monitoring sites 1, 2, 3, 4, 5 and 6 located in the northern part have the highest AQI statistical medians of over 110. However, the monitoring sites 11, 12, and 13 located in the southern part have the lowest AQI statistical medians of less than 100. This spatial distribution characteristic is mainly because of industrial layout. Most industrial districts and people gather at the northern part of Xi'an City. The power plants, industrial manufacturers, automobile exhaust and human activities in the old urban district generate most of the PM2.5 and these zones have become the contamination source. Therefore, the AQI statistical median of PM2.5 is largest in north Xi'an.

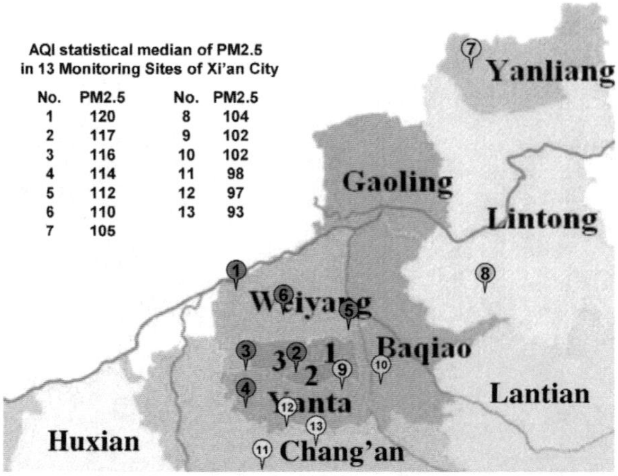

Fig. 4 Locations and AQI statistical medians of 13 monitoring sites in Xi'an

The wind direction also has an important effect on spatial distribution of PM2.5. Fig.5 shows the statistics of wind direction all over the year in Xi'an. As can be seen, the wind direction in Xi'an is mainly east/northeast. Sometimes there is west/southwest wind in winter. However, the southeast/northwest wind is very rare. As a result, little PM2.5 generated in the northern part can be blowed towards the monitoring sites 11, 12 and 13 in southern part. So the AQI statistical medians in these sites are the lowest. In contrast, some PM2.5 generated in the northern part may be blowed towards the monitoring sites 7, 8, 9 and 10 in eastern and northeastern parts when the west/southwest wind comes. So the AQI statistical medians in these sites are lower than that in the northern sites, but higher than that in the southern sites.

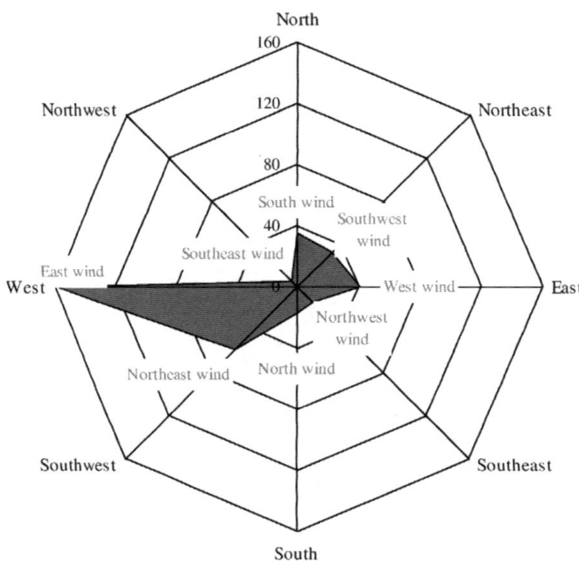

Fig. 5 Statistics of wind direction in Xi'an

3 Conclusion

Based on the AQI statistical data of 13 monitoring sites in Xi'an City since 2013, the paper analyzes the temporal/spatial distribution of PM2.5 and their formation mechanisms. The conclusions are listed as follows:

(1) As for temporal distribution, the AQI values of PM2.5 are highest in winter (December, January, February) and lowest in summer (June, July, August). This is mainly because coal combustion, plant withering, soil exposure and little precipitation increase generation of PM2.5 in winter. Thermal inverse layer and low self-purification ability of air, water and plants affect the diffusion of PM2.5. In addition, the west/southwest and north wind in winter blows PM2.5 from northern source to the whole city. However, in summer, little coal combustion and much precipitation reduce generation of PM2.5. Cross-ventilation and high self-purification

ability of air, water and plants promote the diffusion of PM2.5. The east/northeast wind also blows PM2.5 away from the city

(2) As for spatial distribution, the AQI values of PM2.5 are highest in monitoring sites of the northern city and lowest in that of the southern city. This is mainly because industrial districts and population gather at the northern city and generate most of the PM2.5. Since the southeast/northwest wind is very rare, little PM2.5 can be blowed towards the southern city, but some PM2.5 may be blowed towards the eastern and northeastern city when the west/southwest wind comes.

References:

[1] Kanaya Y, Akimoto H, Wang Z F, et al., Overview of the Mount Tai Experiment (MTX2006) in central East China in June 2006: studies of significant regional air pollution, *Atmospheric Chemistry and Physics*, 2013, 13: 1527-1573.

[2] Cheng Z, Jiang J, Fajardo O, et al., Characteristics and health impacts of particulate matter pollution in China (2001–2011), *Atmospheric Environment*, 2013, 65: 186-194.

[3] Stafoggia M, Cesaroni G, Peters A, et al., Long-term exposure to ambient air pollution and incidence of cerebrovascular events: results from 11 European cohorts within the ESCAPE project, *Environmental Health Perspectives*, 2014, 122(9): 919.

[4] Nawrot T S, Perez L, Künzli N, et al., Public health importance of triggers of myocardial infarction: a comparative risk assessment, *The Lancet*, 2011, 377(9767): 732-740.

[5] Hamra G B, Guha N, Cohen A, et al., Outdoor Particulate Matter Exposure and Lung Cancer: A Systematic Review and Meta-Analysis, *Environmental Health Perspectives*, 2014, 122(9): 906-911.

[6] Pu W W, Zhao X J, Zhang X L, et al., Effect of meteorological factors on PM2.5 during July to September of Beijing, *Procedia Earth and Planetary Science*, 2011, 2: 272-277.

[7] Zhao J, Zhang F, Chen J, et al., Characterization of polycyclic aromatic hydrocarbons and gas/particle partitioning in a coastal city, Xiamen, Southeast China, *Journal of Environmental Sciences*, 2010, 22(7): 1014-1022.

[8] Lippmann M. The 1997 US EPA standards for particulate matter and ozone, *Issues in environmental science and technology*, 1998, 10: 75-100.

[9] Ambient air quality standards, *Standards Press of China*, GB3095-2012, 2012.

[10] Technical Regulation on Ambient Air Quality Index (on trial), *Environmental Standards Press of China*, HJ633-2012, 2012.

[11] Daily Air Quality Index of 13 monitoring sites in Xi'an, *http://www.xianemc.gov.cn/sxmpcp_qt.asp?lb=%D6%CA%C1%BF%C8%D5%B1%A8.*

The effect of biochar on the soil hydrophysical proporties

DUŠAN IGAZ[1], JÁN HORÁK[1], JANA DOMANOVÁ[1], MIRIAM JAROŠOVÁ[1], VILIAM BÁREK[2], PETER HALAJ[2]
[1]Department of Biometeorology and Hydrology
[2]Department of Landscape Engineering
Slovak University of Agriculture
Hospodárska 7, 94976 Nitra, SLOVAKIA
dusan.igaz@uniag.sk, www.spu.sk

Abstract: This study was conducted at the experimental site of SAU-Nitra (Nitra-Malanta) in Nitra region of Slovakia (lat. 48°19´00´´; lon. 18°09´00´´) during March-October, 2014.The main objective of this study was to evaluate the effect of biochar (BC) application on the soil hydrophysical properties (saturated hydraulic conductivity, soil water retention curve). Biochar used for the field experiment was produced from paper fiber sludge and grain husks (1:1 w/w) by pyrolysis at 550°C for 30 minutes. The biochar was incorporated into the soil (0-10 cm) before sowing the spring barley (Hordeum vulgare L.) in spring 2014. The experiment consisted of 3 levels of biochar application 0, 10 and 20 t/ha (BC0, BC10, BC20) with 3 replicates to a loamy soil of 9 experimental plots (6x4m). The soil samples (3 replicates) were taken 53 days after biochar application from each experimental plot from the soil depth of 5-10 cm. Generally, application of biochar increased saturated hydraulic conductivity compared to the control. Application of 20 t/ha of biochar had significant effect on the saturated hydraulic conductivity. However, application of 10 and 20 t/ha of biochar had no significant effect on the soil moisture at different negative pressure water potential ϕ_p (p = 0, -1, -5.5, -20, -55, -100kPa).

Key-words: biochar, saturated hydraulic conductivity, soil water retention curve

1 Introduction

The basic hydrophysical parameters are soil water retention curve (*pF*) and saturated soil hydraulic conductivity (*K*). Soil hydraulic conductivity describes the ability of the soil to conduct water and is a function of soil permeability (K_p), the fluidity of moving liquid in the soil (*f*) and the soil moisture (θ). Consequently, a vicious circle develops; further aggravating climate change, soil degradation, erosion, loss of soil organic matter and nutrient leaching. Therefore, sustainable concepts for increased food production are urgently needed to lower pressure on soil, in order to reduce or prevent the negative environmental impact of intensive agriculture [1].

Biochar (BC) application has received a growing interest as a sustainable technology to improve highly weathered or degraded tropical soils [10], BC can enhance plant growth by improving soil chemical characteristics (i.e., nutrient retention, nutrient availability), soil physical characteristics (i.e., bulk density, water holding capacity, permeability), and soil biological properties, all contributing to an increased crop productivity [7], [8], [10]. Biochar refers to the aromatic carbon materials produced by pyrolysis of biomass (heating in an oxygen-limited environment at temperatures 400-900°C). There have been reported a lot of positive impacts of biochar as a soil amendment such as mitigating climate change [4], through increasing soil carbon storage while decreasing direct greenhouse gas emission and improving soil fertility and crop productivity [4], [5]. BC production is considered to be a more effective way for long-term carbon storage in soil in comparison with the direct soil application of the precursory materials.

BC applied into the soil can affect hydrophysical soil characteristics and the porosity and therefore the water retention in the soil by means of different mechanisms. The first way influence is directly from inside the pores of BC, the next one is by creation of wrapping pores between biochar and the surrounding environment of the soil aggregates and lastly by improving of soil pore persistence due to the increased stability of soil aggregates. BC disposes of high stability in the soil [2] and therefore long-term effects on soil retention properties as well as hydrophysical properties can be expected.

2 Problem Formulation

This study examines the hypothesis that biochar incorporation into the soil will improve soil hydrophysical properties with increase of BC application doses.

2.1 Experimental site and experimental design

Experiment was conducted at the experimental site of SAU-Nitra (Nitra-Malanta), in Nitra region of Slovakia (lat. 48°19´00´´; lon. 18°09´00´´) during March-October, 2014. The soil type is classified as Orthic Luvisol (FAO, 1998). The average annual air temperature was 10.3°C and annual precipitation was 640mm during the studied year [6]. The biochar was incorporated into the soil (0-10 cm) before sowing the spring barley (Hordeum vulgare L.) in spring 2014. The experiment consisted of 3 levels of biochar application 0, 10 and 20 t/ha (BC0, BC10, BC20) with 3 replicates to a loamy soil of 9 experimental plots (6x4m). The soil samples (3 replicates) were taken 53 days after biochar application from each experimental plot from the soil depth of 5-10 cm. The plots were separated by a protection row 0.5 m in width.

2.2 Determination of hydraulic conductivity using the falling head method

A device based on the principle of measurement with variable hydraulic slope (i.e., above the measured soil sample changes the water level during time) was used to determine the hydraulic conductivity in the laboratory condition. Both principles require determination of hydraulic conductivity on undisturbed soil samples taken in sampling soil rollers. Soil samples were taken into non-corrodible metal rollers with a volume of 100 cm^3. The soil sample was saturated by water before measurements were taken. The saturation lasted for 24 hours [3], [6]. The value of the hydraulic conductivity coefficient was calculated according to the equation:

$$K = \frac{L}{t}.\ln\frac{H_2}{H_1} \ (\mathrm{m.s^{-1}}) \qquad (1)$$

where L is the length of roller (cm), t is time of water drop from height H_2 to H_1 (s), H_2 is the initial height of water in the extension piece (another roller whose flow surface is equal to the flow surface of soil sample) (cm) and H_1 is the height of water in the extension piece after the water level dropped (cm).

2.3 Determination of soil water retention curve

Drainage branch of water retention curve was determined by pressure method using the extractor with ceramic pressure plates. Retention curves of the soil sample comprise relation between soil moisture and negative pressure head. Soil moisture was determined at specific water potentials ϕ_p (p = 0, -1, -5.5, -20, -55, -100 kPa). The attention was focused on soil porosity and soil water content at specified pressures during the results evaluation. The water content in the soil samples was then estimated gravimetrically by drying the samples at 105 °C till constant weight.

2.4 Biochar characterization

Biochar used for the field experiment was produced from paper fiber sludge and grain husks (1:1 w/w) (company Sonnenerde, Austria) by pyrolysis at 550°C for 30 minutes in a Pyreg reactor (Pyreg GmbH, Dörth, Germany). Basic chemical and physical properties of the biochar (biochar certificate Nr.1013069001 provided by company Sonnenerde, Austria) are: pH (KCL) 8.8, Ca 57.0 g kg^{-1}, Mg 3.9 g kg^{-1}, K 15.0 g kg^{-1}, Na 0.7 g kg^{-1}, Total C 53.1 g kg^{-1}, Total N 14.0 g kg^{-1}, C/N 37.9.

2.5 Statistical analysis

The statistical processing of the data included the determination of means and standard deviations. One-way analysis of variance (ANNOVA) at $p<0.05$ was performed to examine the effect of biochar at different application rates on soil pH. Significant differences between means of seven replications were identified using the least significant difference test (LSD at $p<0.05$) using STATGRAPHICS Centurion statistical software.

3 Problem Solution

After assessing the effect of biochar on soil saturated hydraulic conductivity 53 days after application in the conditions of Danubian Upland (Slovakia) can be concluded that the highest saturated hydraulic conductivity value was reached after application of BC20 (Fig. 1). The saturated hydraulic conductivity was higher by 14,38 cm.day^{-1} as compared to control and the difference was statistically significant ($p< 0.05$). The similar results were also found in other study [12] where the saturated hydraulic conductivity in clay-rich soil was increased.

Lover application of biochar (BC10) also increased saturated hydraulic conductivity compared to control. With hydraulic conductivity being higher by 3,83 cm.day^{-1} in comparison to control, however the difference wasn't statistically significant (p< 0.05).

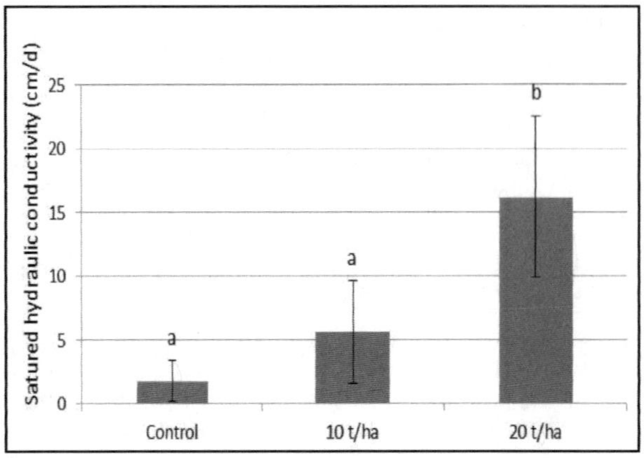

Fig. 1 The effect of biochar application on the saturated hydraulic conductivity. Error bars show standard deviations (n = 7) and the letter above the error bars indicates significant difference evaluated by LSD test ($p < 0.05$).

Table 1 The effect of biochar application on soil moisture at different matric potential ϕ_p (p = 0, -1, -5,5, -20, -55, -100 kPa) and saturated hydraulic conductivity

Pressure potentials	Treatment					
	BC0 Control		BC 10t/ha		BC 20t/ha	
Θ_S (0kPa)	45,4	±2,32 a	48,9	±3,94 b	47,7	±0,92a b
Θ (-1,0kPa)	42,6	±2,86 a	45,5	±3,16 a	42,9	±2,20a
Θ (-5,5kPa)	37,9	±3,70 a	42,5	±5,36 a	38,2	±4,79a
Θ (-20kPa)	31,2	±0,67 a	31	±1,08 a	30,7	±1,01a
Θ (-55kPa)	28,6	±0,79 a	29,2	±1,29 a	28,8	±1,73a
Θ (-100kPa)	26,7	±0,90 a	26,2	±1,58 a	26,5	±1,08a
K (cm/day)	1,78	±1,63 a	5,61	±4,04 a	16,16	±6,29b

All values represent means ± standard deviation (n=7). Different superscript letters represent significant differences between treatments at the $p < 0.05$ level determined by LSD test.

The effect of BC application on the soil hydrophysical properties is showed in Table 1 and Figures 1, 2 and 3. Table 1 shows soil moisture values for variants of biochar application (control, BC10 and BC20). Soil moisture was determined in pressure apparatus at specific pressure potential ϕp (p = 0, -1, -5.5, -20, -55, -100kPa). Table

1 and figures 2 and 3 show that the highest mean soil moisture values were measured at BC10 treatment. Soil moisture increase of this treatment was recorded at pressure potentials ϕ_p - 1,0kPa, -5.5 kPa and -55 kPa, respectively, but without significant differences as compared to control. There were found no differences between the individual treatments at ϕ_p -20kPa and -100kPa.

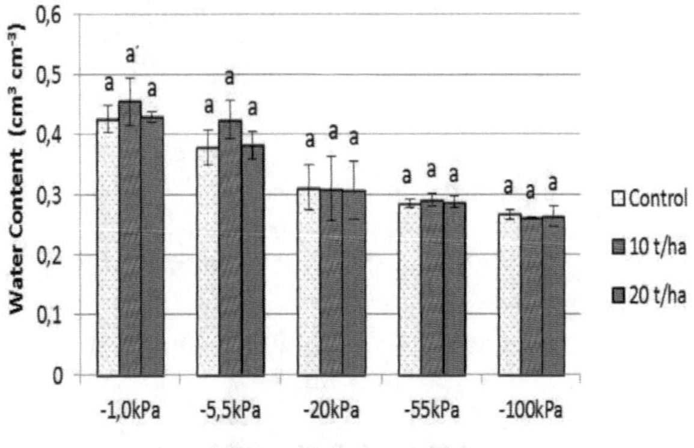

Fig. 2 The effect of biochar application on the soil water retention. Error bars show standard deviations (n=7) and the same letters above the error bars indicate no significant difference between treatments evaluated by LSD test ($p < 0.05$).

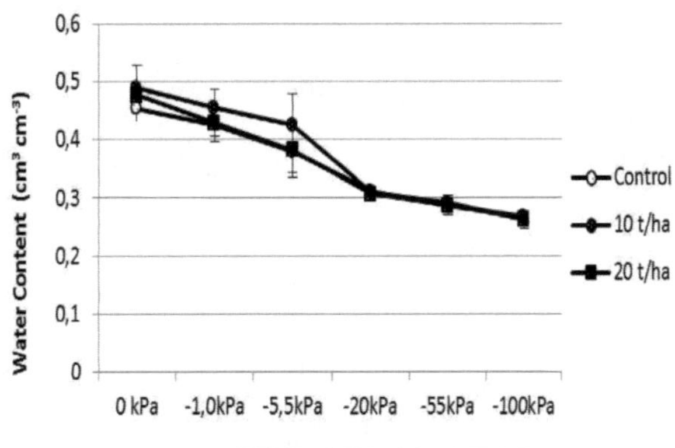

Fig. 3. The effect of biochar application on the soil water content at different negative pressure potentials. Error bars show standard deviations (n=7).

4 Conclusions

This study investigated the effect of different biochar amendments (0, 10, 20 t/ha) on the saturated hydraulic conductivity and the soil moisture at the different water potentials of the Orthic Luvisol. The soil samples (3 replicates) were taken 53 days after biochar application from each experimental plot from the soil depth of 5-10 cm. Application of biochar increased saturated hydraulic conductivity compared to the control. The saturated hydraulic conductivity value after the application of BC10 and BC20 was increased by 3,83 $cm.day^{-1}$, and 14,38 cm.day`1, respectively. However only higher Biochar application rate (20 t/ha) had a significant effect on the saturated hydraulic conductivity. Application of 10 and 20 t/ha of biochar had no significant effect on the soil moisture at different negative pressure water potentials ϕ_p (p = 0, -1, -5.5, -20, -55, -100kPa).

Acknowledgement

This study was supported by Slovak Grant Agency –VEGA, No. 2/0040/12 and Slovak Research and Development Agency under the contract No. APVV-0512-12.

References:

[1] Čimo, J. et all. *Biometeorology*. 2014, Nitra : Slovak University of agriculture in Nitra, 2014. 145 p. ISBN 978-80-552-1219-7

[2] Hardie, M., Clothier, B., Bound, S., Oliver, G., Close, D. 2013, Does biochar influence soil physical properties and soil water availability? In *Plant Soil*. DOI 10,1007/s11104-013-1980-x, Springer, 2013

[3] Igaz, D., Domanová, J., - Tárnik, A. 2014, Influence of selected soil physical parameter (two measurement methods) on soil moisture simulation. *In Ecology & Safety*. ISSN 1313-2563, 2014, vol. 8, no. 1, s. 335-346,

[4] Lehmann, J. 2007 Bio-energy in the black, Front. *Ecologic. Environ*., vol. 5, 2007, pp. 381–387.

[5] Glaser, B., Haumaier, L., Guggenberger, G. & Zech, W. 2001, 'The Terra Preta phenomenon: a model for sustainable agriculture in the humid tropics', *Naturwissenschaften*, vol. 88, 2001, pp. 37-41.

[6] Domanová, J., Igaz, D., Borza, T., Horák, J. 2015, The impact of biochar on hydraulic conductivity of the soil. In *Agrosym 2015* ISBN 978-99976-632-1-4. Agrosym. Sarajevo : Univerzitet u Sarajev, 2015, s. 498.

[7] Hidetoshi A., et al. 2009, Biochar amendment techniques for upland rice production in Northern Laos: 1. Soil physical properties, leaf SPAD and grain yield. In *Field Crops Research* Volume 111, Issues 1–2, 15 March 2009, Pages 81–84 doi:10.1016/j.fcr.2008.10.008

[8] B. Glaser, J. Lehmann, W. Zech. 2002 Ameliorating physical and chemical properties of highly weathered soils in the tropics with biochar a review. In*: Biology and Fertility of Soils*, 35 (2002), pp. 219–230

[9] J. Lehmann, J.P. da Silva Jr., M. Rondon, C. Steiner, T. Nehls, W. Zech, B. Glaser. 2002 Nutrient availability and leaching in an archaeological Anthorosol and a Ferralsol of the Central Amazon basin: fertilizer, manure and charcoal amendments In: *Plant and Soil*, 249 (2002), pp. 343–357

[10] J. Lehmann, M. Rondon. 2006 Bio-char soil management on highly weathered soils in the humid tropics. In: *Biological Approaches to Sustainable Soil Systems*, CRC Press, Boca Raton, FL (2006), pp. 517–530

[11] Y. Ogawa, H. Kato, M. Ishikawa 1992 A simple analytical method for index of soil nitrogen availability by extracting in phosphate buffer solution In:*Japanese Society of Soil Science and Plant Nutrition*, 60 1992, pp. 160–163

[12] Barnes RT, Gallagher ME, Masiello CA, Liu Z, Dugan B 2014 Biochar-Induced Changes in Soil Hydraulic Conductivity and Dissolved Nutrient Fluxes Constrained by Laboratory Experiments. PLoS ONE 9(9): e108340. doi:10.1371/journal.pone.0108340

Tillage systems and fertilization effect on seasonal CO_2 and N_2O emissions for orthic luvisol under spring barley

JÁN HORÁK[1], DUŠAN IGAZ[1], VILIAM BÁREK[2], PETER HALAJ[2]
[1]Department of Biometeorology and Hydrology
[2]Department of Landscape Engineering
Slovak University of Agriculture
Hospodárska 7, 94976 Nitra, SLOVAKIA
jan.horak@uniag.sk, www.spu.sk

Abstract: The objective of this study was to evaluate the effect of conventional tillage (CT) and reduced tillage (RT) combined with unamended control (N0) and addition of nitrogen fertilizers (N1) on soil N_2O and CO_2 emission from cropland during the studied period (April-December in 2012). Weekly N_2O and CO_2 fluxes, soil temperature, and soil water content were monitored during the studied period in western Slovakia. Soil CO_2 emission generally did not differ between conventional tillage compared with reduced tillage for both nitrogen treatments. A significant correlation ($p < 0.05$) between the CO_2 flux and soil temperature in conventionally and reduced tilled plots could lead to the easy conclusion that only the soil temperature (and not the soil moisture) was a controlling factor for fertilized and not fertilized plots, which wasn't our case. N_2O emission decreased in order CT>RT for both fertilization levels N0 and N1. The cumulative N_2O fluxes from CT with N0 treatment (6.4 kg N_2O-N ha^{-1}) were higher by 68 % compared to the RT (3.8 kg N_2O-N ha^{-1}). And cumulative N_2O fluxes from CT with N1 treatment (7.0 kg N_2O-N ha^{-1}) were higher by 66 % compared to the RT (4.2 kg N_2O-N ha^{-1}). In case of N_2O can be concluded that management events play an important role in N_2O emission patterns where fertilization, tillage, harvest combined with rain events and increasing soil temperature had the highest influence.

Key-Words: N_2O, CO_2, conventional tillage, reduced tillage, fertilization, agro-ecosystem

1 Introduction

The greenhouse gas exchange between croplands and atmosphere plays an important role in the global carbon cycle and carbon dioxide concentration in the atmosphere (CO_2).

Conventional tillage (CT) system as a traditional farming management has its advantages and disadvantages. For example advantages of this system is, that soil nutrients are more intensively used by crops, which compensate the costs of expensive mineral fertilizers. CT system completely inverts the soil and has the positive effect on weediness. However there is a growing trend for a conversion of CT system to conservation tillage systems including reduced till (RT) and no-tillage (NT) systems.

With the adoption in the EU of the Kyoto Protocol, there is a considerable interest in agricultural practices that may lead to a net reduction in greenhouse gas emissions [13]. The benefits of the reduced tillage may be at least partly offset by increased emissions of nitrous oxide (N_2O) [30]. There is an uncertainty associated with current estimates of the influence of tillage practice on N_2O emissions. For example some studies reported that N_2O fluxes can be greater under NT system as compared to CT system [4], [22]. Other studies show similar values of N_2O fluxes under both CT and NT systems [17], [29]. On the other there are reported as well higher N_2O emissions under CT system as compared to NT system [15].

Reduced tillage is regarded as one of the most effective agricultural practices to reduce CO_2 emission and sequester atmospheric C in the soil [18]. Studies showed that decreased tillage intensity reduces soil disturbance and microbial activities, thus reducing CO_2 emission [20]. On the other hand, increased tillage intensity showed increased CO_2 emission through increasing aeration due to increased soil disturbance [27] and to physical degassing of dissolved CO_2 from the soil solution [25]. The effect of tillage on CO_2 emission, however, was short-lived (<24 hours). Studies on effect of tillage on soil CO_2 emission are mostly limited to the period between tillage and crop establishment [21] or short-term CO_2 flux [26]. Large number of studies on CO_2 emission during growing season is from grass or forest, but not from cropland [23], [14]. The effect of different tillage practices during crop growing season is not well known. There are no such studies from the agricultural used soils of Slovakia at all.

Implementation and spreading of conservation systems including reduced tillage in condition of Slovak Republic is limited by large acreage of not suitable compacted soils or partly compacted soils. But there is also an absence of detailed regionalization of suitable soils for this conservation systems as well as absence of long term field research in different ecological conditions. Estimates of agricultural GHG emissions are needed to develop economically efficient as well as effective policies in mitigating and reducing GHG emissions in farming systems.

The objective of this study was to evaluate the effect of conventional tillage (CT) and reduced tillage (RT) combined with unamended control (N0) and addition of nitrogen fertilizers (N1) on seasonal N_2O and CO_2 emissions from cropland under the spring barley during the studied period (April-December in 2012) on a long-term tillage field experiment initiated in 1994. Other factors such as soil water content and temperature were monitored to see their role (magnitude of influence) in soil CO_2 and N_2O emissions.

2 Problem Formulation

2.1 Field site and experimental design

This study was conducted at the Experimental site of the Department of Plant Production of SAU-Nitra, in Nitra region of Slovakia (lat. 48°19´00´´; lon. 18°09´00´´) during April-December 2012 on a long-term tillage field experiment initiated in 1994. The soil type was classified as Orthic Luvisol (FAO 1998) containing 360.4 g kg^{-1} of sand, 488.3 g kg^{-1} of silt and 151.3 g kg^{-1} of clay [31]. The average annual air temperature was 11.4°C and annual precipitation was 492mm during the studied year.

The experiment compared two tillage methods (conventional tillage-CT and reduced tillage-RT) combined with unamended control (N0) and addition of nitrogen fertilizers (N1). Experiment was arranged in a split plot design with tillage as the main plots and the treatments N0 and N1 as the sub-plots with three replicates. CT consisted of tillage to 22 – 25 cm and RT consisted of disking to 10-12 cm depth both applied each fall, followed by harrow cultivator to 10 cm depth each spring before seeding. The spring barley (Kangoo variety) was seeded on 16 March 2012 at a rate of 4 500 000 seeds ha^{-1}. The doses of fertilizers were calculated by balance method and were applied to the soil twice throughout the season. First application (CT 50 kg N ha^{-1} and RT 40 kg N ha^{-1}) was applied on 11 April and second (CT and RT 7 kg N ha^{-1}) on 23 July 2012. Both CT and RT plots were disked (10-12 cm) after harvest (12 July, 2012) on 24 July 2012 and crop residues were incorporated into the soil.

2.2 Measurements of GHGs (soil N_2O and CO_2), water content and temperature

The soil surface of N_2O and CO_2 fluxes were measured weekly (between 10 A.M and 12 A.M.) in all treatments during and after the crop growing season from April-December 2012 using closed chamber technique. The metal collar frame was inserted 10 cm deep into the soil in every plot and left undisturbed until harvest/disking occasion. On every gas sampling, the chamber (30 cm in diameter and 25 cm in height) were water sealed onto bottom collars and gas samples (20 mL) were collected through tube fittings (sealed with septum) at 0, 30 and 60 min after chamber deployment using an air-tight syringe (Hamilton) and transferred to pre-evacuated 12 mL glass vials (Labco Exetainer). Gas samples were analyzed for N_2O and CO_2 using a gas chromatograph (GC-2010 Plus Shimadzu). N_2O was detected using an electron capture detector (ECD) and CO_2 was detected using a thermal conductivity detector (TCD). Gas fluxes were calculated from the changes in headspace concentration during the chamber closure using a linear approach. Cumulative total CO_2 and N_2O emissions were calculated by interpolating the emissions between each sampling day. Soil water content at 0-10 cm depth (gravimetric method) and soil temperature at 5 cm depth (Volcraft DET3R thermometer) was measured at each gas sampling event. Water-filled pore space (WFPS) was calculated using the bulk density (measured monthly by the core method) in each plot assuming a mineral particle density of 2.65 g cm^{-3}.

2.3 Statistical analyses

The statistical processing of the data included the determination of means, Pearson coefficients of correlation at the confidence level of $p<0.05$. The significance of the differences between the means was evaluated using one-way ANOVA at $p<0.05$.

3 Problem Solution

3.1 Soil CO_2 emission

Long-term CO_2 emission in the 2012 growing season from conventional (CT) and reduced tillage (RT) plots combined with and without N fertilizer application (N1, N0) followed similar patterns initially, increasing towards the spring barley filling stage in early summer and decreasing towards the autumn/winter season (Fig. 1). In general, the average daily CO_2 emissions were higher in both tillage treatments (CT and RT) by 19 and 18% in case of fertilized plots (Tab. 1) as compared to unfertilized CT and RT plots. This suggests the fact that N fertilization increased CO_2 emission when soil water content is adequate. Limited or no increase in CO_2 emission with increased N fertilization rate had been reported by several researchers [28], [32].

There wasn't significant difference ($p < 0.05$) in soil CO_2 between conventional tillage and reduced tillage for both fertilization levels. The average CO_2 emissions from the soil were almost exactly the same in the CTN0 and RTN0, reaching values of 87.2 and 87.1 kg ha^{-1} day^{-1}, respectively (Tab. 1). The same was found in case of fertilized plots, when CO_2 emissions reached almost the same values of 103.5 and 102.9 kg ha^{-1} day^{-1} for CTN1 and RTN1, respectively. The highest CO_2 emissions were observed in mid-June (Fig. 1.) for both tillage treatments. Increase in soil moisture during this period due to precipitation events (Fig. 3.) together with high soil temperature (Fig. 2.) can be associated with the highest CO_2 emission. There is an exponential increase in soil CO_2 fluxes with respect to temperature increases observed for biological systems [24]. A rapid increase in soil CO_2 fluxes within minutes after precipitation and rewetting of dry soils has been observed in a variety of ecosystems [16], [1].The rapid release of CO_2 is due to the infiltration of rainfall water into the soil pores, which replaces highly concentrated CO_2 air, resulting in degassing [2]. Degasing is not considered to be soil respiration as it releases the stored CO_2 in soil from past microbial and root respiration. The CO_2 flux of as much as 300 kg ha^{-1} day^{-1} following heavy rain in dry soil in northern Great Plains had been known [8]. Seasonal variation in the CO_2 emissions under long term field measurements are attributed to changes in soil temperature under agricultural soils irrespective of the tillage systems [3], [10], [1] i.e. higher emissions measured in the summer season in comparison to the autumn/winter season due to higher C mineralization rates with increase in temperature [19].

Our study also showed clear decline in CO_2 emissions just after barley harvest and then followed by second highest CO_2 peak just after incorporation of crop residues into the soil for both tillage treatments whether they were fertilized or not. Second emission peak patterns can be attributed to the combined effects of soil temperature, water availability and tillage. Finally, rainfall, high temperatures, and the crop residue available for decomposition embedded in the soil could result in a soil environment which enhanced high microbial activity, thereby increasing C mineralization and CO_2 flux during the summer on CTN0, RTN0, CTN1 and RTN1. On quantifying the total emissions during the studied period (April-December 2012) we found that they were 21.8 and 25.9 t ha^{-1} in CTN0 and RTN0 and 21.5 and 25.7 t ha^{-1} in CTN1 and RTN1, respectively (Tab. 1). Total emission didn't indicated an increase of soil CO_2 flux at the end of studied period when soils were tilled with conventional tillage as compared to reduced tillage for both nitrogen treatments (N0,N1).

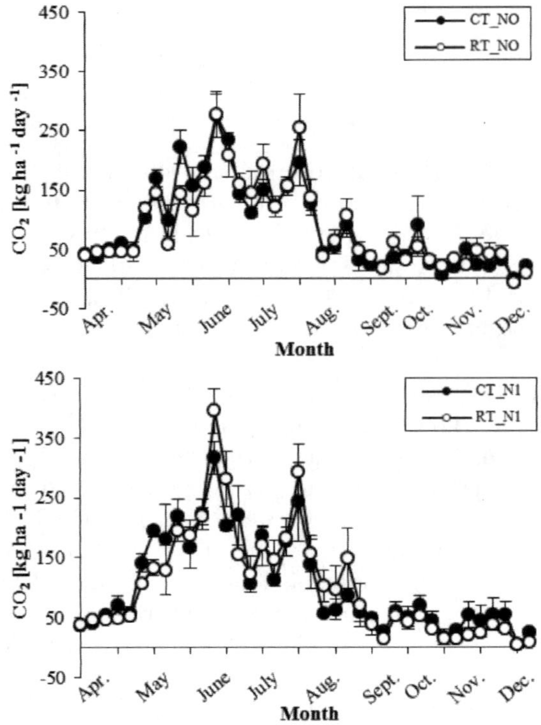

Fig. 1 Soil CO_2 emission under conventional (CT) and reduced tillage (RT) combined with and without N fertilizer application (N0, N1)

Table 1 Mean, maximum, minimum and total cumulative CO_2 emissions during the studied period

Soil tillage and N level	CO_2 emissions			
	Mean	Maximum	Minimum	Total
	kg ha^{-1} day^{-1}			t ha^{-1}
CTN0	87,2	275,4	-1,6	21,8
RTN0	87,1	276,9	-7,0	21,5
CTN1	103,5	316,5	5,1	25,9
RTN1	102,9	394,9	5,0	25,7

3.2 Relationship between CO_2 and soil temperature and soil water content

The results showed that CO_2 emissions for both tillage treatments combined with and without nitrogen fertilizers generally followed the trend of soil temperature (Fig. 1 and Fig. 2). The Pearson correlation coefficient between the CO_2 flux and soil temperature during the studied period was significant ($p < 0.05$) for both conventional (r = 0.52 for CTN0; r = 0.49 for CTN1) and reduced tillage (r = 0.60 for RTN0; r = 0.58 for RTN1) whether the plots were fertilized or not.

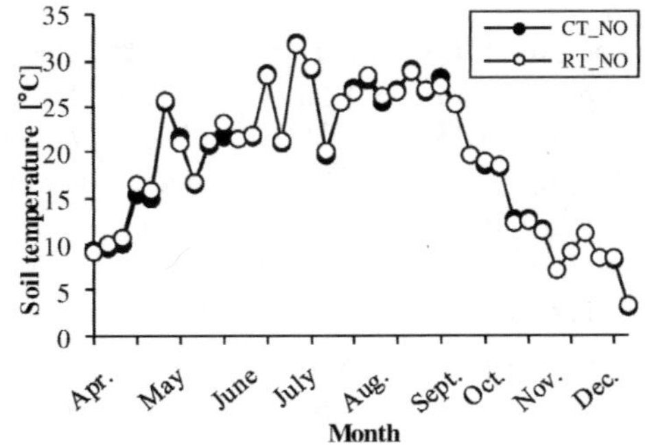

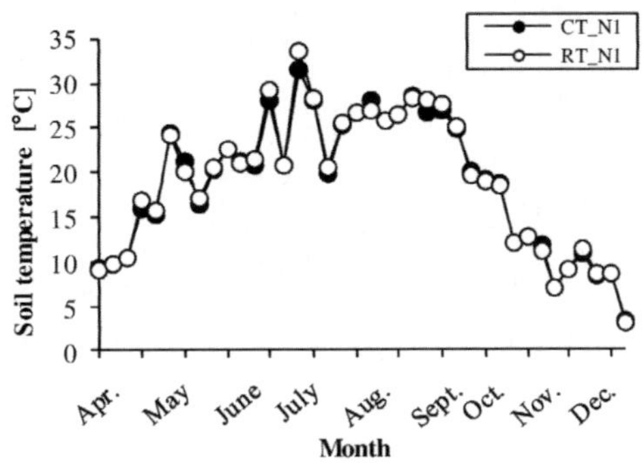

Fig. 2 Soil temperature under conventional (CT) and reduced tillage (RT) combined with and without N fertilizer application (N0, N1)

On the other hand, there wasn't found any significant correlation between the CO_2 flux and soil water content for both conventional and reduced tillage systems combined with and without N. High CO_2 flux usually occurs in the summer when soil temperature is higher and soil water content and substrate C availability is adequate, while low emission occurs in the winter when soil biological activity is minimal due to near-freezing soil temperature [12], [5], [6].

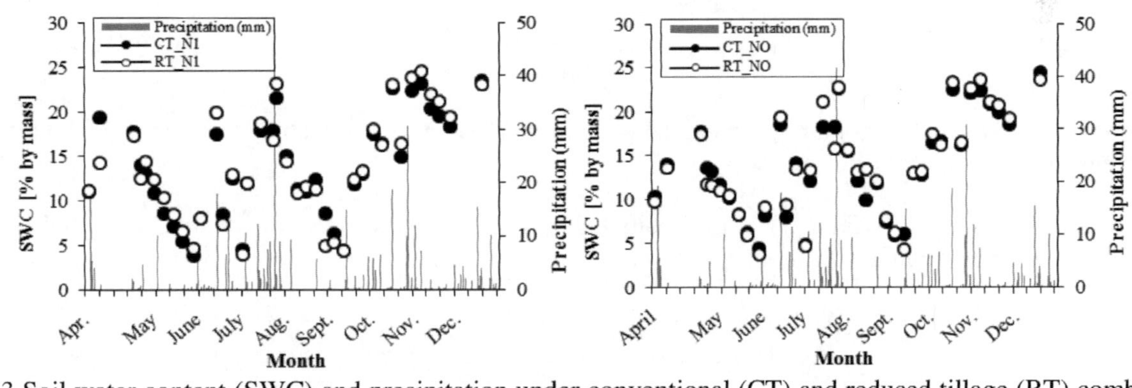

Fig. 3 Soil water content (SWC) and precipitation under conventional (CT) and reduced tillage (RT) combined with (N1) and without (N0) N fertilizer application

3.3 Soil N_2O emission

In general, daily mean N_2O emissions were slightly higher in both tillage treatments (CT and RT) in case of fertilized plots (Tab. 2) as compared to unfertilized CT and RT plots. This suggests the fact that N fertilization increased N_2O emission when soil water content is adequate which was also found in case of soil CO_2 emissions. Daily N_2O fluxes were low or was found uptake for the first three months (April-June) with the highest values being measured during July-August with emissions peaks related to the harvest, disking, second fertilization (only for N1 treatment) combined with rain events and increasing soil temperature (Fig. 4) for all treatments (CTNO, RTNO, CTN1, RTN1). There was also measured one emission peak at the end of

November, however only in case when the plots were conventionally tilled. This peaks were attributed to tillage (depth 22 – 25 cm) which was done just week before on plots with CTN0 and CTN1 treatments.
There wasn't significant difference ($p < 0.05$) in soil N_2O fluxes between conventional tillage and reduced tillage for both, not fertilized and fertilized plots. The differences between the treatments are better to be seen by calculating cumulated fluxes. Table 2 shows the comparison of the cumulated N_2O fluxes over the studied period between the two tillage systems and nitrogen treatments. N_2O emission decreased in order CT>RT for both fertilization levels N0 and N1. The cumulative N_2O fluxes from CT with N0 treatment (6.4 kg N_2O-N ha^{-1}) were higher by 68 % compared to the RT (3.8 kg N_2O-N ha^{-1}). And cumulative N_2O fluxes from CT with N1 treatment (7.0 kg N_2O-N ha^{-1}) were higher by 66 % compared to the RT (4.2 kg N_2O-N ha^{-1}).

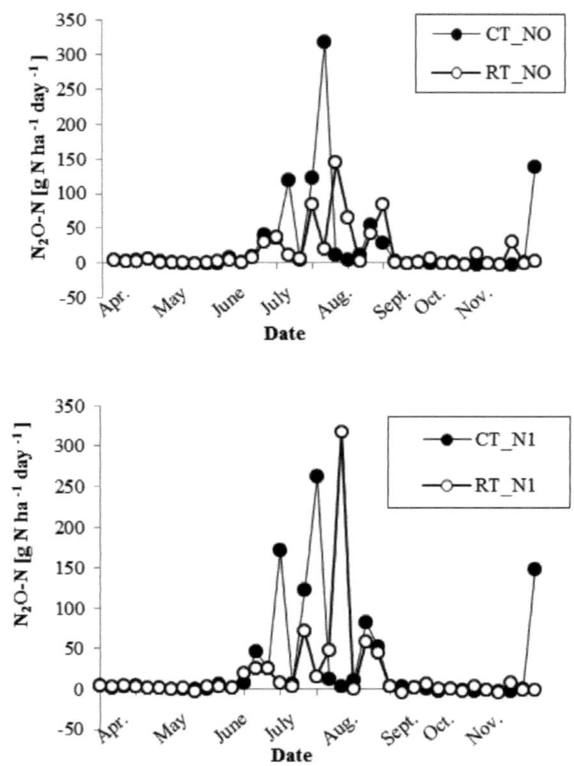

Fig. 4 Daily mean N_2O fluxes during the studied period for different tillage treatments combined with nitrogen treatments (CT and RT with N0 and N1)

Table 2 Mean, maximum, minimum and total cumulative N_2O emissions under conventional (CT) and reduced tillage (RT) combined with and without N fertilizer application (N0, N1)

Soil tillage and N level	N_2O emissions			
	Mean	Maximum	Minimum	Total
	g ha^{-1} day^{-1}			kg ha^{-1}
CTN0	23.8	318.2	-3.7	6.4
RTN0	15.6	143.7	-3.1	3.8
CTN1	25.3	261.6	-3.7	7.0
RTN1	17.4	317.2	-5.0	4.2

3.4 Relationship between N_2O and WFPS and soil temperature

Soil water-filled pore space (WFPS) depends not only on the rainfall but is also affected by the soil physical properties and mechanical impacts on the soil. The highest correlation between WFPS and N_2O was found only at conventionally tilled plots with no nitrogen application (N0) ($R = 0.26$, p =0.120) as well as at fertilized plots

(N1) ($R = 0.17$, p =0.296) even the soil WFPS was very low during the studied period varying mostly between 30-60% (Table 3). However, both correlations were insignificant. There wasn't found any correlation between WFPS and N_2O at reduced tilled plots with both nitrogen treatments (N0, N1). In order to have maximum N_2O production rates, soil WFPS must generally be between 70 and 90 % WFPS [7]. The greatest fluxes were found when WFPS exceeded 45 % (Fig. 4 and Fig. 5). Daily N_2O fluxes from the soil could be very low even after the fertilization (available N in the soil) as it was found at the beginning of our study after the first application of fertilizer (11 April), indicating that other parameters for example temperature and soil wetness [5] were controlling N_2O production in the soil. This was found in the study [10] which observed that application of N-fertilizer by itself did not always enhance N_2O emission and that weather conditions at the time of fertilizer application influenced N_2O emissions substantially.
Soil temperature is one of the key parameters affecting N_2O emission from N-fertilized agricultural soils but only if the parameters, such as WFPS and soil available N, are not limiting the processes of N_2O production in the soil [7]. High daily N_2O emission during the studied period from all treatments were mostly found when the temperature of the top 5cm layer was higher than 20°C. There was found higher correlation between the temperature and N_2O from the all treatments as it was found between WFPS and N_2O. The highest and significant correlation between soil temperature and N_2O was found at RTN0 treatment ($R = 0.45$, p =0.04) (Table 3).

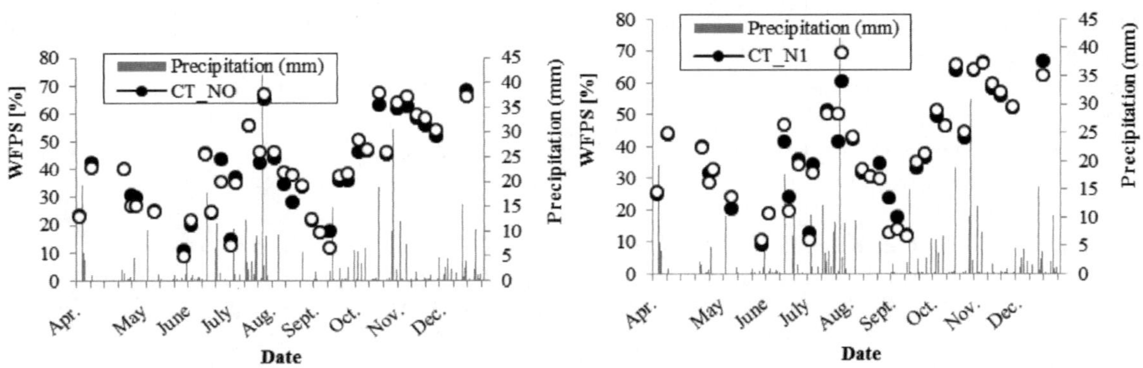

Fig. 5 Measured daily precipitation, water filled pore space (WFPS) for different tillage treatments combined with nitrogen treatments (CT and RT with N0 and N1)

Table 3 Pearson correlation between N_2O fluxes and % WFPS and soil temperature for different tillage treatments combined with nitrogen treatments (CT and RT with N0 and N1). The numbers in bold are correlation coefficients and the numbers in italics are the p values (insignificant at $p > 0.05$)

N_2O emissions	% WFPS	Soil Temp.
CT N0	**0.26**	**0.27**
	0.120	*0.098*
CT N1	**0.17**	**0.30**
	0.296	*0.065*
RT N0	**0.00**	**0.45**
	0.992	*0.049*
RT N1	**-0.07**	**0.31**
	0.655	*0.061*

4 Conclusions

Results of this study showed that the average CO_2 emissions were higher in conventional tillage and reduced tillage when the plots were fertilized as compared to unfertilized plots. It can be also concluded that soil CO_2 emission didn't significantly differ between conventional tillage as compared to reduced tillage for both nitrogen treatments. A significant correlation ($p < 0.05$) between the CO_2 flux and soil temperature in conventionally and reduced tilled plots could lead to the easy conclusion that only the soil temperature (and not

the soil moisture) was a controlling factor for fertilized and not fertilized plots, which wasn't our case. It can be concluded that increase in soil water content due to precipitation events together with high soil temperature were associated with the highest CO_2 emissions and smaller CO_2 fluxes but still high, when the soil was disturbed by crop residues incorporation after harvest.
In case of N_2O can be concluded that management events play an important role in N_2O emission patterns. Fertilization, tillage, harvest combined with rain events and increasing soil temperature had the highest influence. It could be also suggested that since the conventionally tilled soils were better aerated, the soil aeration and diffusivity might played the role in higher emissions from CT.

Acknowledgement
This study was supported by Slovak Grant Agency –VEGA, No. 2/0040/12 and Slovak Research and Development Agency under the contract No. APVV-0512-12.

References:

[1] Almagro, M., Lopez, J., Querejeta, JI. & Martinez-Mena, M., Temperature dependence of soil CO2 efflux is strongly modulated by seasonal patterns of moisture availability in a Mediterranean ecosystem, *Soil Biology & Biochemistry*, Vol. 41, No. 3, 2009, pp. 594-605.
[2] Alvaro-Fuentes, J., Lopez, MV., Arrue, JL. & Cantero-Martinez, C., Management effects on soil carbon dioxide fluxes under semiarid Mediterranean conditions, *Soil Science Society of America Journal*, Vol. 72, No. 1, 2008, pp. 194-200.
[3] Aslam, T., Choudhary, MA. & Saggar, S., Influence of land-use management on CO2 emissions from a silt loam soil in New Zealand, *Agriculture Ecosystems & Environment*, Vol. 77, No. 3, 2000, pp. 257-262.
[4] Baggs, E.M., Stevenson, M., Pihlatie, M., Regar, A., Cook, H. & Cadish, G., Nitrous oxide emissions following application of residues and fertilizer under zero and conventional tillage, *Plant Soil*, Vol. 254, 2003, pp. 361-370.
[5] Bajracharya, RM., Lal, R. & Kimble, JM., Erosion effects on carbon dioxide concentration and carbon flux from an Ohio alfisol, *Soil Sci. Soc. Am. J.*, Vol. 64, 2000a, pp. 694-700.
[6] Bajracharya, RM., Lal, R. & Kimble, JM., Diurnal and seasonal CO2-C flux from soil as related to erosion phases in central Ohio, *Soil Sci. Soc. Am. J.*, Vol. 64, 2000b, pp. 286-293.
[7] Conen, F., Dobbie, K.E. &Smith, K.A., Predicting N_2O emission from agricultural land through related soil parameters, *Glob change Biol,* Vol. 6, 2000, pp. 417-426.
[8] Curtin, D., Wang, H., Selles, F., McConkey, BG. & Campbell, CA., Tillage effects on carbon fluxes in continuous wheat and fallow-wheat rotations, *Soil Sci. Soc. Am. J.*, Vol. 64, 2000, pp. 2080-2086.
[9] Dobbie, K.E., McTaggart, I.P. & Smith, K.A., Nitrous oxide emissions from intensive agricultural systems: variations between crops and seasons, key driving variables, and mean emission factors, *J.Geophys.Res-Atmos*. Vol. 104, 1999, pp. 26891-26899.
[10] Elder, JW. & Lal, R., Tillage effects on gaseous emissions from an intensively farmed organic soil in North Central Ohio, *Soil & Tillage Research*, Vol. 98, No.1, pp. 45-55.
[11] Flessa, H. & Dorsch, P., Seasonal variation on N_2O and CH_4 fluxes in differently managed arable soils in southern Germany, *J. Geophys. Res.*, Vol. 100, 1995, pp. 23115-23124.
[12] Follett, RF., CRP and microbial biomass Dynamics in temperate climates', in R Lal (ed.), Management of carbon sequestration in soil, *CRC Press. Boca Raton*, 1997, pp. 305-322.
[13] Freibauer, A., Rounsevell, M.D.A., Smith, P. & Verhagen, J., Carbon sequestration in European agricultural soils, *Geoderma*, Vol. 122, 2004, pp. 1-23.
[14] Frank, AB., Liebig, MA. & Hanson, JD., Soil carbon dioxide fluxes in northern semiarid grassland, *Soil biology and biochemistry*, Vol. 34, pp. 1235-1241.
[15] Jacinthe, P. & Dick, W., Soil management and nitrous oxide emissions from cultivated fields in southern Ohio, *Soil Tillage Res.*, Vol. 41, 1997, pp. 221-235.
[16] Jenerette, GD., Scott, RL. & Huxman, TE., Whole ecosystem metabolic pulses following precipitation events, *Functional Ecology*, Vol. 22, No. 5, 2008, pp. 924-930.
[17] Kessavalou, A., Doran, J.W., Mosier, A.R. & Drijber, R.A., Greenhouse gas fluxes following tillage and wetting in a wheat-fallow cropping system, *J. Environ Qual*, Vol. 27, 1998, pp. 1105-1116.
[18] Kern, JS. &Johnson, MG., Conservation tillage impacts on national soil and atmospheric carbon levels, *Soil Sci. Soc, Am. J.*, Vol. 57, 1993, pp. 200-210.

[19] Kirschbaum, MUF., The temperature-dependence of soil organic-matter decomposition, and the effect of global warming on soil organic-C storage, *Soil Biology & Biochemistry*, Vol. 27, No. 6, 1993, pp. 753-760.

[20] Lal, R. & Kimble, JM., Conservation tillage for carbon sequestration, *Nutr. Cycling Agroecosyst.* Vol. 49, 1997, pp. 243-253.

[21] La Scala, N. Jr., Lopes, A., Marques J. Jr. & Pereira, GT., Carbon dioxide emissions after application of tillage systems for a dark red latosol in southern Brazil, *Soil Till.Res.*, Vol. 62, 2001, pp. 163-166.

[22] MacKenzie, A.F., Fan, M.X. & Cadrin, F., Nitrous oxide emissions as affected by tillage, corn-soybean-alfalfa rotations and nitrogen fertilization, *Can. J. Soil Sci.*, Vol. 77, 1997, pp. 145-152.

[23] Maljanen, M., Hytonen, J. & Martikainen, PJ., Fluxes of N_2O, CH_4 and CO_2 on afforested boreal agricultural soils, *Plant and soil*, Vol. 231, 2001, pp. 113-121.

[24] O'Connell, AM., Microbial decomposition (respiration) of litter in eucalypt forests of south-western Australia - an empirical-model based on laboratory incubations, *Soil Biology & Biochemistry*, Vol. 22, No. 2, 1990, pp. 153-160.

[25] Reicosky, DC. & Lindstrom, MJ., Fall tillage method: Effect on short-term carbon dioxide flux from soil, *Agron. J.*, Vol. 85, 1993, pp. 1237-1243.

[26] Reicosky, DC., Reeves, DW., Prior, SA., Runion, GB., Rogers, HH. & Raper, RL., Effects of residue management and controlled traffic on carbon dioxide and water loss, *Soil Till. Res.*, Vol. 52, 1999, pp. 153-165.

[27] Roberts, WP. & Chan, KY., Tillage-induced increases in carbon dioxide loss from soil, *Soil Tillage Res.*, Vol. 17, 1990, pp. 143-151.

[28] Rochette, P. & Gregorich, EG., Dynamics of soil microbial biomass C, soluble organic C, and CO_2 evolution after threee years of manure application, *Can. J. Soil Sci.*, Vol. 78, 1998, pp. 283-290.

[29] Robertson, G.P., Paul, E.A. & Harwood, R.R., Greenhouse gases in intensive agriculture: Contributions of individual gases to the radiative forcing of the atmosphere, *Science*, Vol. 289, 2000, pp. 1922-1925.

[30] Six, J., Ogle, S.M., Breidt, F.J., Conant, R.T., Mosier, A.R. & Paustian, K., The potential to mitigate global warming with no-tillage management is only realized when practised in the long term, *Glob Change Biol.*, Vol. 10, 2004, pp. 155-160.

[31] Šimanský, V., Tobiašová, E. & Chlpík, J., Soil tillage and fertilization of Orthic Luvisol and their influence on chemical properties, soil structure stability and carbon distribution in water-stable macro-aggregates, *Soil Tillage Res.*, Vol. 100, 2008, pp. 125-132.

[32] Wagai, R., Brye, KR., Gower, ST., Norman, JM. & Bundy, LG., Land use and environmental factors influencing soil surface CO_2 flux and microbial biomass in natural and managed ecosystems in southern Wisconsin, *Soil Biol. Biochem.*, Vol. 30, 1998, pp. 1501-1509.

Treatment of Pickles Wastewater By A2/O Filter Bed Technology

JINHAI WU[a,b], LIEZHUAN ZHANG[c], GANG GAO[a,b,*]
[a] College of Food Science, Shanxi Normal University, Linfen Shanxi 041000, P. R. China
[b]Lab of Central and Local Governments, Linfen Shanxi 041000, P. R. China
[c]Linfen People's Hospital, Linfen Shanxi 041000, P. R. China
517613124@qq.com

Abstract: The start-up step of A2/O biological filter bed device was studied by gradually increasing the water salinity after the pre-biofilm formation, as well as the salinity adaptability of the device. The results showed that the biological filter bed system was successfully established after around 50 days with the concentrations of CODcr, suspended solid and ammonia stabilized at 82.7, 35.4 and 13.5 mg.L-1, respectively, which accords with the 1st grade of national effluent standard. Furthermore, with a high salinity of 5%, the water quality, including CODcr, suspended solid and ammonia, after treatment stills meets the 2nd grade of national effluent standard, which indicates a satisfactory and reliable salinity adaptability of the device design.

Key words: salinity, A^2/O filter bed, COD_{cr}, suspended solid, ammonia.

1 Introduction

The discharge of waste water with a high salinity is extremely to the aquatic organisms as well as the drinking water resource and the industrial and agricultural production water use. The strong inhibition effect of mineral salt to the general microbiology makes the appropriate biological treatment of wastewater less efficient and lots of research has been done in this domain in order to improve the performance of wastewater treatment including the new materials and techniques study and development during decades (Chiam and Sarbatly, 2011) as well as different types of aerobic reactors, SBR and biological contact reactor (Ju, 2002; Nong et al., 2008). Considering the specific acquirements for water with high salinity, numerous technologies are used to improve the treatment efficiency by, such as, solar (Todd, 1988), elodea nuttallii (Bishop and Eighmy, 1989), water hyacinths (Tchobanoglous et al., 1989), biological activated carbon (Huang et al., 2011), ozone (Schroeder et al., 2011), water hyacinth (Eichhornia crassopes) and lettuce (Pistia stratiotes) (Akinbile and Yusoff, 2011). Fiber based methods have been widely applied (Kuhn, 2008; Pfeiffer et al., 2008; Kuhn et al., 2010; Chang and Cajucom, 2011). However, the performance of available fiber materials is still need to be improved according to Kuhn (2008). Chiam and Sarbatly (2011) indicated that the new membrane-based and biofilm techniques with fiber material could be used to achieve higher treatment efficiency with relatively lower cost.

The degradation of organic pollutants is realized by the metabolism of halophilic and halotolerant bacteria in the treatment of high salinity wastewater, the difficulty of which lies in the construction of microbial ecosystem structure. A gradual domestication method was used by Zhou (2007) and a microbial system was successfully constructed in the treatment of wastewater from preserved pickle production process (7% of salinity). In this work, an A2/O biological filter bed process was carried out for the treatment of wastewater from a pickles factory in ZiBo with a salinity of 3%. The start-up step has been done by using a pre-biofilm formation on the filter bed packing material, followed by a progressive increase of salinity, and the salinity tolerance characteristics of the devices has also been studied.

2 Experimental Materials and Methods

2.1 Experimental device

An A2/O process of the filter bed was used in this work and the designed inlet flow rate is 20 L.d-1. The integrated experimental device was made of plexiglas and shown in Fig. 1.

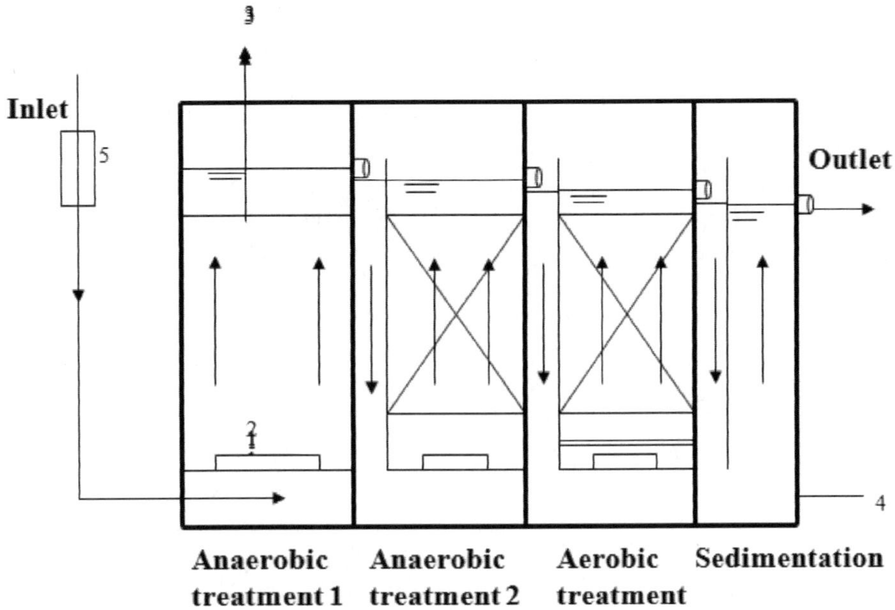

Fig. 1 Schematic diagram of the A2/O experimental device (1. Distributor; 2, Aeration pipe; 3, Sampling point; 4, Sludge discharge; 5, Flowmeter).

The integrated process design including two anaerobic and an aerobic treatment as well as a sedimentation zone uses fewer materials, which is more ecological. The effective volume of anaerobic and aerobic zone is different and a bottom-up low pattern is used in each zone. The wastewater flows uniformly through the packing material by using the distributors, which ensures the full contact with the biofilm. The aeration pipe and sludge discharge pipe are installed in the aerobic treatment and sedimentation zone, respectively. A PVC porous corrugated material is used as the packing material in the anaerobic and aerobic treatment zone, of which the surface area of unit volume is 40 m2.m-3. The parameters of device are shown below in Table 1.

Tab.1 Parameters of experimental device.

Name of zone	Length×width×height (m)	Effective volume (L)	Hydraulic Retention Time (h)
Anaerobic treatment 1	0.27×0.32×0.45	30	36
Anaerobic treatment 2	0.22×0.32×0.45	24	28
Aerobic treatment	0.16×0.32×0.45	17	20
Sedimentation	0.07×0.32×0.45	7.5	9

2.2 Analytical Methods and Experimental Procedure

The samples are collected once each two days from the three sampling points and the inlet/outlet of wastewater. Each of the samples is measured twice, including the CODcr, ammonia, measured using GB determination of China[00], and dissolved oxygen (DO), measured by an DO instrument (Thermo Electron Corporation 850A). The mean value of these two measurements for each parameter is used to further analysis.

The wastewater with a high salinity from a pickles production company in Zibo City is used in this study and the raw water has to be pretreated with a mesh filter of 5 mm in order to prevent the eventual clogging in the filter bed by the pickles residual materials. The raw water quality is shown below in Table 2.

Tab.2 Raw water quality

	pH	salinity	COD_{Cr} ($mg.L^{-1}$)	suspended solid ($mg.L^{-1}$)	ammonia ($mg.L^{-1}$)
Raw water	5 - 8	3 %	1000～1200	220～280	80～110

This wastewater, containing the soluble proteins, vitamins, sugars, acids and other organic pollutants, with a BOD/COD higher than 50% is suitable for the biochemical treatment. However, the high concentration of salinity and ammonia cause great difficulties to the treatment process.

The pre-biofilm produced by urban wastewater on the filter bed, then followed by a progressive increase of salinity is used to accelerate the start-up step considering that only few types of halophilic bacteria are capable of growth and reproduction in a hardly biodegradable wastewater with a high salinity. During the experiment, the concentration of dissolved oxygen in the aeration treatment zone is kept at 4 mg.L-1.

Anaerobic and aerobic sludge from this company is used in the pre-biofilm formation step and inoculated in the experimental device, which is filled with urban wastewater. The CODcr of the wastewater is adjusted to 1100 mg.L-1 by adding glucose solution and the ammonia concentration augments to 90 mg.L-1 by adding the ammonium chloride. The anaerobic and aerobic microorganism is able to grow under these conditions and the biofilm progress can be successfully realized on the surface of filter bed. The growth condition of biofilm is under surveillance and observation during 14 days and the wastewater of different salinity with the sequence from 0.5%, 1%, 2%, to 3% is then used, which is produced by mixing glucose, pickles and urban wastewater. A flow rate of 20 L.d-1 is adopted and an adaption stage of 10 days for each salinity is used in order to investigate the treatment efficiency of each zone in the device.

3 Results and discussion

The COD_{cr} removal is widely used as an essential indicator in the biochemical wastewater treatment process, which is shown below in Fig. 2.

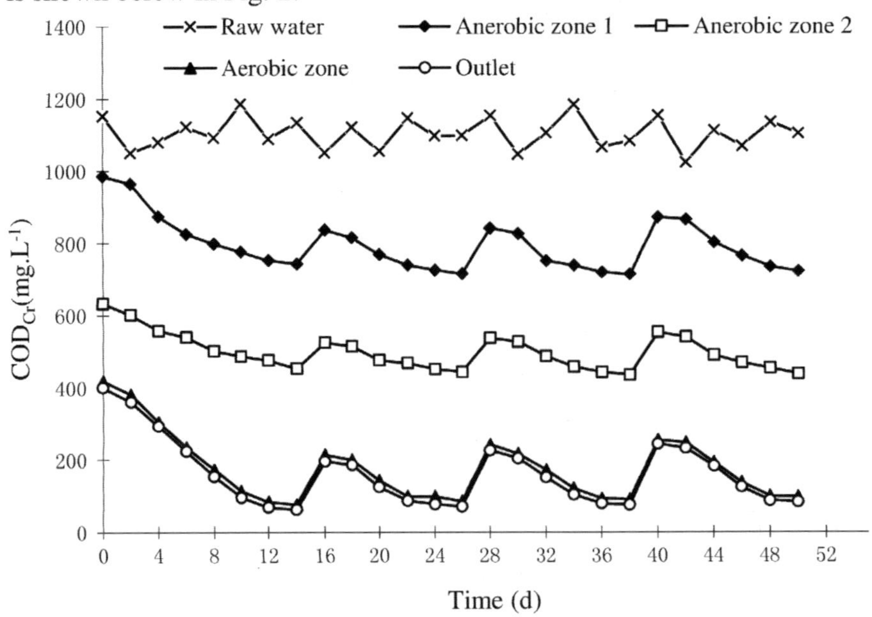

Fig. 2. Variance of COD_{cr} concentration in different treatment zones.

As indicated in Fig. 2. the CODcr concentration is reduced sequentially through different treatment zones of A2/O filter bed, in which 50% of CODcr reduction is realized by anaerobic treatment zones. Considering the energy conservation of anaerobic treatment in this process, the operating costs could be greatly decreased in the industrial installations. Then the pretreated wastewater can be further treated by the aerobic biofilm, passing through the aerobic zone and the CODcr concentration reduction achieves around 92.5%, which accords with the 1st grade of national effluent standard.

During the start-up process, the urban wastewater without salinity is firstly used with the inoculated sludge with a rapid adaptive phase, leading to a quick formation of anaerobic and aerobic biofilm on the filter bed, in which a yellow-brown biofilm can be observed on the aerobic filter bed after 3 days and a black biofilm attached on the anaerobic filter bed after 8 days. The CODcr concentration is stabilized at 62.1 mg.L-1 around 14 days at the outlet of device and the pre-biofilm formation step is completed.

Pickles wastewater contains certain residual impurities such as fine residues and cellulose, leading to a relatively high suspended solid concentration. The SS concentration variance at the outlet of each treatment zone is shown in the Fig. 3.

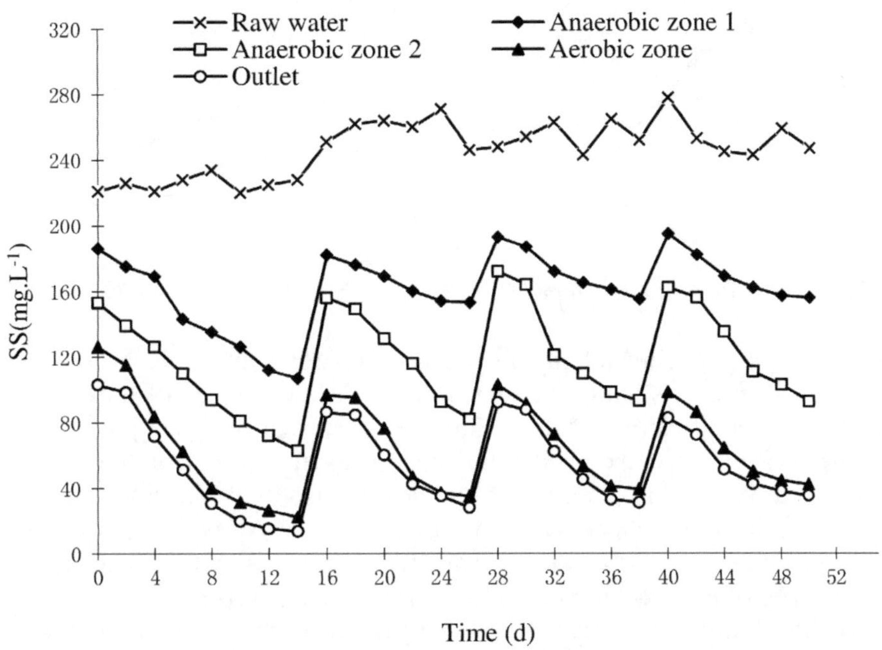

Fig. 3. Variance of SS concentration in different treatment zones.

As seen from Fig. 3. a high removal efficiency of SS concentration could be obtained after passing through A2/O filter bed, even with a salinity of 3% and the SS concentration of 35.4 mg.L-1 at the outlet of device also reaches the 1st grade of national effluent standard, which is because of the filtration and adsorption characteristics of filter bed. The suspended solids are firstly trapped in the filter bed, then adsorbed and degraded by the biofilm on the packing bed. A slight difference between the SS removal efficiency of urban wastewater and pickles wastewater is notified (Fig. 3.), indicating that the degradation of suspended solids in the pickles wastewater is more difficult compared with that in the urban wastewater.

From the results above, the wastewater of different salinity is used (0.5%, 1%, 2%, 3%) after the pre-biofilm formation start-up step and the CODcr , SS and ammonia concentrations at the outlet of each treatment zone increase significantly at the beginning of the salinity change, then decrease gradually since then, which is caused by the inhibited metabolism of the biofilm due to the increased salinity, causing the death and inactivation of microorganism under this high salinity environment. A new ecosystem needs to be created, which demands a certain period of time representing the adaptive phase of biofilm, leading to a gradually enhanced adsorption and filtration capacity and accordingly an increased CODcr, SS and ammonia removal efficiencies. Furthermore, the inactivated microbial flows with the wastewater, which also leads to a higher SS concentration at the outlet of each treatment zone.

The effluent of ammonia pollutants can lead to the eutrophication, causing significant harm to the environment by consuming the dissolved oxygen in the water, which consequently affects the respiration of aquatic organisms. The ammonia concentration variance is shown in the Fig. 4.

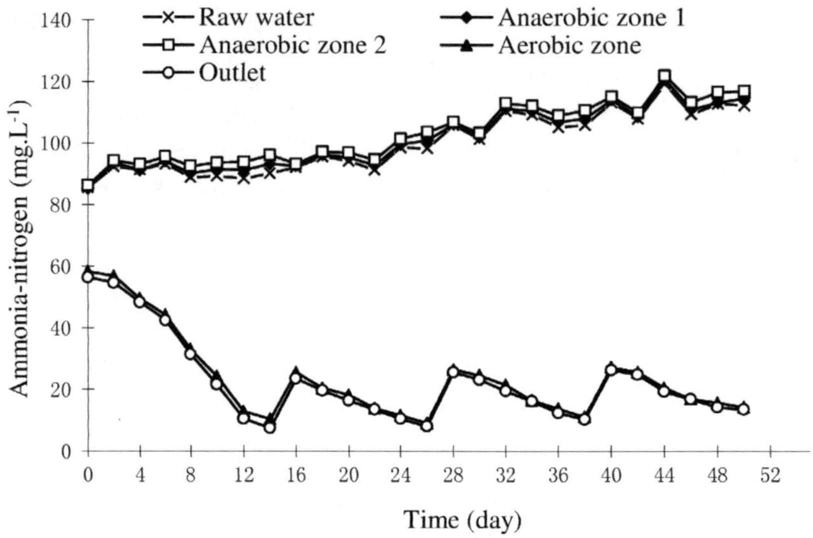

Fig. 4. Variance of Ammonia concentration in different treatment zones.

The removal of ammonia occurs mainly in the aerobic treatment zone as shown in Fig. 4. The ammonia concentration increases slightly passing through the two stages of anaerobic treatment zone, which is due to the ammonification of the nitrogenous substances like protein by the anaerobic microorganism in the biofilm, while the nitrification occurs under aerobic conditions by the nitrobacteria and the ammonia can be converted into nitrate. The high ammonia concentration at the outlet of device at the beginning of pre-biofilm formation start-up step is mainly caused by the lack of nitrobacteria in the initial inoculated sludge and the nitrification gradually strengthens with the development of aerobic biofilm where the quantity of nitrobacteria keeps increasing. The adaptive phase is also observed with the wastewater of different salinity, which represents the establishment process of the new ecosystem in the biofilm.

All the results discussed above indicate that a certain period of time is necessary for the device in order to adapt the sudden change of salinity affecting the metabolism and the evolution of the microorganism in the biofilm, which could be further verified by the higher salinity (3.5%, 4%, 5%) by adding sodium chloride in the raw water. The effluent quality at the outlet of each treatment zone is shown in Fig. 5.

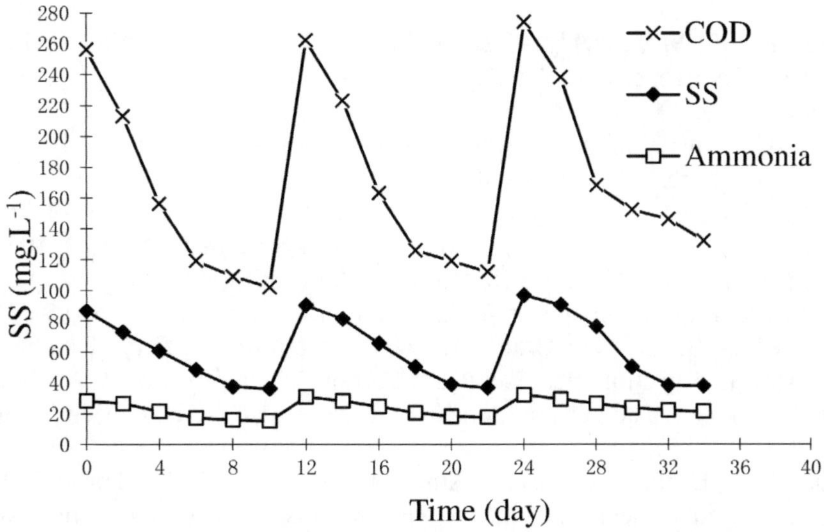

Fig. 5. Variation of effluent characteristics at the outlet of device.

The similar tendencies are observed using the wastewater with the higher salinity (3.5%, 4% and 5%), compared with the raw water with the lower salinity (0.5%, 1%, 2% and 3%). An adaptive phase to the new environment is inevitable for the biofilm after increasing the salinity of the raw water and the treatment efficiency deteriorates with the increasing salinity, leading the pollutants removal less efficient, which is verified by the increased CODcr, suspended solid and ammonia concentration at the outlet of device (Fig. 5). However, with the highest salinity of 5%, the CODcr, suspended solid and ammonia concentration is respectively 132, 37.8 and 21.3 mg.L-1, which accords still with the 2nd grade of national effluent standard. For the industrial scale of pickles wastewater treatment process, the hydraulic retention time increases dramatically by expanding the volume of each treatment zone leading a better treatment efficiency, which could be coupled with other physical and chemical process for the further treatment.

4 Conclusion

The pre-biofilm formation is successfully realized for the start-up step using urban wastewater with the increasing salinity in the A2/O filer bed device. The CODcr, suspended sludge and ammonia concentration of the wastewater with the salinity 3% is respectively 82.7, 35.4 and 13.5 mg.L-1, which accords well with the 1st grade of national effluent standard. The two-stage anaerobic treatment zone the A2/O filter bed device is the mainly effective area in this process and the removal of CODcr and suspended sludge after the anaerobic treatment zone is both above 50%. The operating costs can be greatly reduced considering energy conservation of anaerobic treatment in the process. With the further increased salinity from 3.5% to 5%, the treatment efficiency of the biofilm on the filter bed is slightly inhibited. However, the concentration of CODcr, suspended sludge and ammonia still meets the 2nd grade of national effluent standard, which indicates a satisfactory and reliable salinity adaptability of the A2/O filter bed device.

Acknowledgments

This work has been supported by Research Project Supported by Shanxi Scholarship Council of China and Agricultural Research project 20130311029-1 from Department of agriculture of Shanxi Province, China.

References:

1. [1] Akinbile, C., Yusoff, M.S., 2011. Assessing Water Hyacinth (Eichhornia crassopes) and Lettuce (Pistia stratiotes) Effectiveness in Aquaculture Wastewater Treatment. International Journal of Phytoremediation, 14(3), 201-211.
2. [2] Bishop, P.L., Eighmy, T.T., 1989. Aquatic wastewater treatment using Elodea nuttallii. Water Pollution Control Federation,61(5), 641-648.
3. [3] Chang, C.Y., Cajucom, S.S., 2011. Feasibility Study of Fish Farm Effluent Treatment by Sequencing Batch Membrane Bioreactor. Journal of Water Sustainability, 1(1), 103-112.
4. [4] Chiam C, Sarbatly R., 2011. Purification of Aquacultural Water: Conventional and New Membrane-based Techniques. Separation & Purification Reviews 40(2): 126-160.
5. [5] Huang XT, Chen B, Liu W, Qian YJ, 2011. Study on Tertiary Treatment of Circulating Aquaculture Wastewater by Biological Activated Carbon. Technology of Water Treatment, 2011-06.
6. [6] Ju, H., 2002. Pickles wastewater treatment by SBR reactor. Guizhou environmental protection technology 2(31): 9-22.
7. [7] Kuhn DD, 2008. Enhancing Shrimp Aquaculture Sustainability Through Water Reuse and Biological Treatment. PhD Thesis, Virginia Polytechnic Institute and State University.
8. [8] Kuhn DD, Lawrence AL, Boradman GD, Patnaik S, Marsh L, Flick Jr GJ, 2010. Evaluation of two types of bioflocs derived from biological treatment of fish effluent as feed ingredients for Pacific white shrimp, Litopenaeus vannamei. Aquaculture 303(1-4): 28-33.
9. [9] Nong, S.M., Li, H.D., Zhang, S.Z., 2008. New development on the treatment of wastewater with high salinity. Jiangsu Environmental technology 21(3): 72-76.
10. [10] Pfeiffer TJ, Osborn A, Davis M, 2008. Particle sieve analysis for determining solids removal efficiency of water treatment components in a recirculating aquaculture system. Aquacultural Engineering 39(1): 24-29.
11. [11] Schroeder JP, Croot PL, Dewitz BV, Waller U, Hanel R, 2011. Potential and limitations of ozone for the removal of ammonia, nitrite, and yellow substances in marine recirculating aquaculture systems. Aquacultural Engineering 45(1): 35-41.
12. [12] Tchobanoglous G, Maitski FK, Thomason K, 1989. Evolution and performance of city of San Diego pilot scale aquatic wastewater treatment system using water hyacinths. Journal of Water Pollution Control Federation 61: 11-12.
13. [13] Todd J, 1988. Solar aquatic wastewater treatment. Biocycle 29(2): 38-40.
14. [14] Zhou, J., Wu, Y.T., Long, T.R., 2007. The microbial ecosystem construction in the treatment of wastewater from preserved pickle production process with high salinity. China water supply 23(15): 17-20.

AUTHOR INDEX